High-Operating-Temperature Infrared Photodetectors

High-Operating-Temperature Infrared Photodetectors

Jozef Piotrowski

Antoni Rogalski

SPIE
PRESS

Bellingham, Washington USA

Library of Congress Cataloging-in-Publication Data

Piotrowski, Józef.
  High-operating-temperature infrared photodetectors / Jozef Piotrowski and Antoni Rogalski.
        p. cm. -- (Press monograph ; 169)
  Includes bibliographical references and index.
  ISBN 978-0-8194-6535-1
  1. Optical detectors. 2. Infrared detectors.   I. Rogalski, Antoni. II. Title.

TK8360.O67P56 2007
681' .25—dc22

                                        2007000344

Published by

SPIE
P.O. Box 10
Bellingham, Washington 98227-0010 USA
Phone: +1 360 676 3290
Fax: +1 360 647 1445
Email: spie@spie.org
Web: http://spie.org

The content of this book reflects the work and thought of the author(s).
Every effort has been made to publish reliable and accurate information herein, but the publisher is not responsible for the validity of the information or for any outcomes resulting from reliance thereon.

Printed in the United States of America.

Contents

Chapter 3   Materials Used for Intrinsic Photodetectors / 33

Chapter 4   Intrinsic Photodetectors / 75

Chapter 7   Photoelectromagnetic, Magnetoconcentration, and Dember IR Detectors / 179

Chapter 8   Lead Salt Photodetectors / 197

Chapter 9   Alternative Uncooled Long-Wavelength IR Photodetectors / 211

Chapter 10   Final Remarks / 233

Acronyms and Abbreviations

A1	Auger 1
A7	Auger 7
AR	antireflection
APD	avalanche photodiode
CBD	chemical bath deposition
DAG	direct alloy growth
D^*	detectivity
DARPA	Defense Advanced Research Projects Agency (U.S.)
DLHJ	double-layer heterojunction
EMI	electromagnetic interference
FOV	field of view
FPA	focal plane array
G-R	generation-recombination
HWE	hot wall epitaxy
HOT	high operating temperature
IMP	interdiffused multilayer process
IR	infrared
ISOVPE	isothermal vapor phase epitaxy
JESS	junction-enhanced semiconductor structure
LN	liquid nitrogen
LPE	liquid phase epitaxy
LWIR	long-wavelength infrared
MBE	molecular beam epitaxy
ML	monolayer
MOCVD	metalorganic chemical vapor deposition
MTBF	mean time before failure
MWIR	middle-wavelength infrared
NEP	noise equivalent power
NETD	noise equivalent temperature difference
PACE	Producible Alternative to CdTe for Epitaxy
PEM	photoelectromagnetic
PC	photoconductive
PV	photovoltaic
QWIP	quantum well infrared photodetector
RA	normalized resistivity

$R_\nu A$	normalized responsivity
RIE	reactive ion etching
ROIC	readout integrated circuit
rms	root mean square
SL	superlattice
SLS	strained layer superlattice
SNR	signal-to-noise ratio
SR	Shockley-Read
TE	thermoelectric
THM	traveling heater method
UV	ultraviolet
VPE	vapor phase epitaxy

Preface

Cryogenic cooling of detectors has always been the burden of sensitive infrared (IR) systems, particularly those operating in the middle-wavelength (MWIR) and long-wavelength (LWIR) range of the IR spectrum. Many efforts have been made to develop imaging IR systems that would not require cryogenic cooling. Until the 1990s—despite numerous research initiatives and the attractions of ambient-temperature operation and low-cost potential—room-temperature IR detector technology enjoyed only limited success in competition with cooled photon detectors for thermal imaging applications. Only the pyroelectric vidicon received much attention, because it was hoped that it could be made practical for some applications. However, the advent of the staring focal plane array (FPA) at the beginning of the 1980s marked the arrival of devices that would someday make uncooled systems practical for many commercial applications.

Throughout the 1980s and early 1990s, many companies in the USA, especially Texas Instruments and Honeywell's research laboratories, developed devices based on various thermal detection principles. By the early 1990s, good imagery was being demonstrated with 320×240 arrays. This success resulted in a decision in the mid-1990s by the U.S. Defense Advanced Research Projects Agency (DARPA) to redirect its support of HgCdTe to uncooled technologies. The goal was to create producible arrays that gave useful performance without the burden of fast ($f/1$) LWIR optics.

Infrared detectors with limited cooling have obvious advantages, including the elimination of power-consuming cryogenics; a reduction in size, weight, and cost; and greater reliability (an increase in the useful life and mean time to failure). The number of applications potentially affected by near room temperature IR detector technology is widespread, including military applications such as battlefield sensors; submunition seekers; surveillance, marine vision, and firefighting devices; hand-held imagers; and general "walking around" helmet-mounted sights. This technology also has widespread civilian applications in areas such as thermography, process control, sensitive heterodyne detection, fast pyrometry, Fourier and laser spectrophotometry, imaging interferometry, laser technology and metrology, long-wavelength optical communication, new types of gas analyzers, imaging spectrophotometers, thermal wave nondestructive material testing, and many others.

The room-temperature operation of thermal detectors makes them lightweight, rugged, reliable, and convenient to use. However, their performance is modest, and they suffer from slow response. Because they are nonselective detectors, their imaging systems contain very broadband optics, which provide impressive sensitivity at a short range in good atmospheres.

During the last decade we have observed a revolutionary emergence of FPAs based on thermal detectors. Their slow speed of response is no limitation for a staring system that covers the whole field of view without mechanical scanning, and their moderate performance can be compensated by a large number of elements in the array.

At present, full TV-compatible arrays are used, but they cannot be expected to replace the high-performance cryogenically cooled arrays without a scientific breakthrough. A response much shorter than that achievable with thermal detectors is required for many applications. Thermal detectors seem to be unsuitable for the next generation of IR thermal imaging systems, which are moving toward faster frame rates and multispectral operation. A response time much shorter than that achievable with thermal detectors is required for many nonimaging applications.

Photon detectors (photodetectors) make it possible to achieve both high sensitivity and fast response. The common belief that IR photodetectors must be cooled to achieve a high sensitivity is substantiated by the huge and noisy thermal generation in their small threshold energy optical transitions. The need for cooling is a major limitation of photodetectors and inhibits their more widespread application in IR techniques. Affordable high-performance IR imaging cameras require cost-effective IR detectors that operate without cooling, or at least at temperatures compatible with long-life, low-power, and low-cost coolers. Since cooling requirements add considerably to the cost, weight, power consumption, and inconvenience of an IR system, it is highly desirable to eliminate or reduce the cooling requirements. Recent considerations of the fundamental detector mechanisms suggest that, in principle, near-perfect detection can be achieved in the MWIR and LWIR ranges without the need for cryogenic cooling.*

This book is devoted to high-operating-temperature (HOT) IR photodetectors. It presents approaches, materials, and devices that eliminate the cooling requirements of IR photodetectors operating in the middle- and long-wavelength ranges of the IR spectrum. This text is based mainly on the authors' experiences in developing and fabricating near room temperature HgCdTe detectors at Vigo Systems Ltd. and at the Institute of Applied Physics Military University of Technology (both in Warsaw, Poland).

The main approaches to minimizing thermal generation in the active element of a detector without sacrificing quantum efficiency include:

- Optimization of the material bandgap and doping;
- Reduction of the detector volume using optical concentrators, backside reflectors, and optical resonant cavities;

---

*See, e.g., J. Piotrowski in *Infrared Photon Detectors*, A. Rogalski (Ed.), 391–494, SPIE Press, Bellingham, WA (1995); C. T. Elliott, N. T. Gordon, and A. M. White, *Appl. Phys. Lett.* **74**, 2881–2883 (1999); and M. A. Kinch, *J. Electr. Mater.* **29**, 809–817 (2000).

- Suppression of Auger thermal generation with nonequilibrium depletion of the semiconductor; and
- Use of new natural and engineered semiconductors with reduced thermal generation.

This text also discusses solutions to other specific problems of high-temperature detection, such as poor collection efficiency due to a short diffusion length, the Johnson-Nyquist noise of parasitic impedances, and interfacing of very low-resistance devices to electronics.

We also consider different types of IR photodetectors, especially photoconductors, photoelectromagnetic (PEM) detectors, Dember effect detectors, and photodiodes, with major emphasis on devices based on HgCdTe alloys. Special attention is given to optimization of the devices for room-temperature operation and the new detector designs that are necessary to solve specific problems of high-temperature operation. Additional topics include an approach proposed by British workers to reduce photodetector cooling requirements based on a nonequilibrium mode of operation, and alternative material systems to the current market-dominant HgCdTe, such as a number of II-VI and III-V semiconductor systems (HgZnTe, HgMnTe, InAsSb) and some artificial narrow-gap semiconductors based on type-II superlattices (InAs-GaInSb).

This book was written for those who desire a comprehensive analysis of the latest developments in HOT IR photodetector technology and a basic insight into the fundamental processes that are important in room-temperature detector operation. Special efforts are focused on the physical limits of detector performance and a performance comparison of different types of detectors. The book is suitable for graduate students in physics and engineering who have received a basic preparation in modern solid state physics and electronic circuits. This book will also be of interest to individuals who work with aerospace sensors and systems, remote sensing, thermal imaging, military imaging, optical telecommunications, IR spectroscopy, and lidar. To satisfy all these needs, each chapter first discusses the principles needed to understand the chapter topic as well as some historical background before presenting the reader with the most recent information available. For those currently in the field, this book can be used as a collection of useful data, as a guide to literature in the field, and as an overview of topics in the field. The book also could be used as a reference for participants of educational short courses, such as those organized by SPIE.

The book starts with two overview chapters. Chapter 1 describes figures of merit, which characterize IR thermal imagers and their cooling requirements. Chapter 2 is an overview of the fundamental limitations to IR photodetector performance imposed by the statistical nature of the generation and recombination process in semiconductor material. Radiometric considerations are also included. In this chapter we try to establish the ultimate theoretical sensitivity limit that can be expected for a detector operating at a given temperature. The model presented in Chapter 2 is applicable to any class of photodetector.

Chapter 3 describes the properties of material systems used to fabricate HOT IR photodetectors. Although the HgCdTe ternary alloy currently holds the dominant market position, a number of II-VI and III-V semiconductor systems are described as alternatives. Chapter 4 presents the general theory of intrinsic detectors with an emphasis on electrical and optical properties, together with the generation-recombination mechanisms that directly influence the performance of IR detectors at near room temperature. The next two chapters describe specific approaches to the most popular HgCdTe detectors: photoconductors and photodiodes. In contrast to photoconductors, photodiodes, with their very low-power dissipation, can be assembled in 2D arrays containing a very large ($>10^6$) number of elements, limited only by existing technologies. At present, photodiodes are the most promising devices for uncooled operation, and significant efforts are directed toward improving the fabrication of multiple heterojunction and Auger-suppressed devices.

Other than photoconductors and photodiodes, three other junctionless devices are used for uncooled IR photodetectors: photoelectromagnetic (PEM) detectors, magnetoconcentration detectors, and Dember effect detectors. The technologies and performances of these three devices are described in Chapter 7. Chapter 8 is devoted to lead salt photodetectors, which were brought to the manufacturing stage of development during World War II. More than 60 years later, low-cost PbS and PbSe polycrystalline photoconductors remain the choice for many applications in the 1–3 μm and 3–5 μm spectral regions. The objective of Chapter 9 is to present the status of alternatives to the current market-dominant HgCdTe detectors. Detectors fabricated from a number of II-VI and III-V semiconductor systems such as HgZnTe, HgMnTe, InAsSb, and type-II InAs/GaInSb superlattices are presented. Chapter 10 presents our final remarks.

The authors have benefited from the kind cooperation of many scientists who are actively working in narrow-gap semiconductor detectors. The preparation of this book was aided by many informative and stimulating discussions between the authors and their colleagues at Vigo System S.A. and the Institute of Applied Physics, Military University of Technology in Warsaw, Poland. These colleagues provided many illustrations and practical results. Special thanks are also extended to L. Faraone, C. Musca, J. Dell, and J. Antoszewski of the Microelectronics Research Group, University of Western Australia, for numerous discussions and their help with manuscript preparation. Thanks also go to SPIE Press, especially Margaret Thayer for her cooperation and care in publishing this text.

Jozef Piotrowski and Antoni Rogalski
February 2007

Chapter 1

Introduction

1.1 General Remarks

Sensors that are used to detect optical radiation are usually confined to use in two types of detectors: thermal detectors and photon detectors. Thermal detectors sense the heat generated by the absorbed radiation, so their operation is a two-step process: conversion of the radiation energy into heat, followed by conversion of the heat energy into the energy of an electrical signal. The incident radiation is absorbed to change the material temperature, and the resultant change in some physical property is used to generate an electrical output.

Despite this two-step operation, thermal detectors are relatively simple devices (see Fig. 1.1) that operate primarily at ambient temperature. In general, a thermal detector is suspended on lags that are connected to a heat sink. The signal does not depend upon the photonic nature of the incident radiation. Thus, thermal effects are generally wavelength independent; the signal depends on the radiant power (or its rate of change) but not on its spectral content, assuming that the mechanism

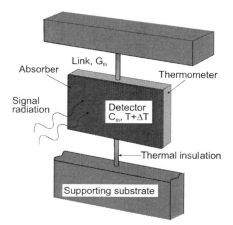

Figure 1.1 The simplest representation of the thermal detector. The detector is represented by a thermal capacitance C_{th} coupled via a thermal conductance G_{th} to a heat sink at a constant temperature T. ΔT is the temperature difference due to the optical signal.

responsible for the absorption of the radiation is itself wavelength independent (see Fig. 1.2).

Three approaches have found the greatest utility in infrared (IR) technology, namely bolometers, pyroelectric effects, and thermoelectric effects. In pyroelectric detectors, a change in the internal electrical polarization is measured, whereas in the case of thermistor bolometers, a change in the electrical resistance is measured. Because they are not selective, thermal detectors can be used in an extremely wide range of the electromagnetic spectrum, from x ray to ultraviolet (UV), visible, IR, and microwave. This is the case for any radiation in which the energy can be readily converted to heat. They have found widespread use in low-cost applications that do not require superior performance and high speed. Their room- temperature operation makes them lightweight, rugged, reliable, and convenient to use. However, they are characterized by modest sensitivity and suffer from a slow response (because heating and cooling of a detector element is a relatively slow process).

Until the 1990s, thermal detectors were considerably less exploited in commercial and military systems in comparison with photon detectors. The reason for this disparity is that thermal detectors were popularly believed to be rather slow and insensitive compared to photon detectors, and useless in scanned thermal imagers. As a result, the worldwide effort to develop thermal detectors was extremely small relative to that of photon detectors. However, during the last decade there has been a revolutionary emergence of focal plane arrays based on thermal detectors.[1] The slow speed of response is no limitation for a staring system covering the whole field of view without mechanical scanning, and thermal detectors' moderate sensitivity can be compensated by a large number of elements in two-dimensional (2D) electronically scanned arrays. With large arrays of thermal detectors, the best values of temperature resolution below 0.05 K can be reached because effective noise bandwidths less than 100 Hz can be achieved. A high sensitivity is also expected by imagers based on superconductor high-temperature bolometers and mechanical cantilever "bimaterial" detectors. Mid-range 10^8 cm Hz$^{1/2}$/W detectivity is typical

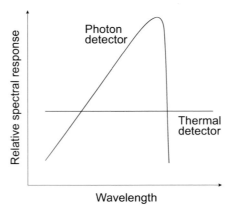

Figure 1.2 Relative spectral response for a photon detector and a thermal detector.

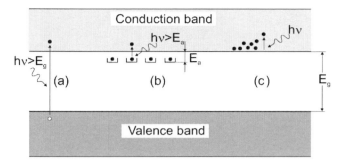

Figure 1.3 Fundamental optical excitation processes in semiconductors: (a) intrinsic absorption, (b) extrinsic absorption, and (c) free-carrier absorption.

for the present microbolometer arrays, while the theoretical limit for radiation heat transfer noise is 1.8×10^{10} cm Hz$^{1/2}$/W at room temperature.

With the second class of IR detectors, photon detectors (photodetectors), the radiation is absorbed within the material by interaction with electrons (see Fig. 1.3) that are either bound to lattice or impurity atoms or are free electrons. The observed electrical output signal results from the changed electronic energy distribution. Photon detectors show a selective wavelength dependence of response per unit of incident radiation power (Fig. 1.2). They exhibit both perfect signal-to-noise performance and a very fast response. But to achieve this, the present photon detectors require cryogenic cooling.

The class of photon detectors is further subdivided into different types based on the nature of the detector's interaction, as shown in Table 1.1. The most important types are intrinsic detectors, extrinsic detectors, photoemissive (metal silicide Schottky barriers) detectors, and quantum well detectors. Depending on how the electric or magnetic fields are developed, there are various modes of operation such as photoconductive, photovoltaic, photoelectromagnetic (PEM), and photoemissive. Each material system can be used for different modes of operation.

A common belief is that an IR photodetector must be cooled to achieve a high sensitivity. The detection of long-wave infrared (LWIR) radiation, which is characterized by low photon energy, requires the electron transitions to be free-charge carriers of energy lower than the photon energy. Therefore, at near room temperatures, the thermal energy kT is comparable to the transition energy. The direct consequence of this is a very high rate of thermal generation by the charge carriers. The statistical nature of this process generates signal noise. As a result, long-wavelength detectors become very noisy when operated at near room temperature. Cooling is a direct, straightforward, and very efficient way to suppress the thermal generation of charge carriers, while at the same time being a very impractical method because it adds considerably to the cost, weight, power consumption, and inconvenience of an IR system. The need for cooling is a major limitation of photodetectors and inhibits the more widespread application of IR technology. Affordable high-performance IR imaging cameras require cost-effective IR detectors that operate without cooling, or at least at temperatures compatible with long-life, low-

Table 1.1 Comparison of IR detectors.

Detector type			Advantages	Disadvantages
Thermal (thermopile, bolometers, pyroelectric)			Light, rugged, reliable, low cost Room-temperature operation	Low detectivity at high frequency Slow response (on the order of ms)
Photon	Intrinsic	IV–VI (PbS, PbSe, PbSnTe)	Easy to prepare Stable materials	Very high thermal expansion coefficient Large permittivity
		II–VI (HgCdTe)	Easy bandgap tailoring Well-developed theory and exper. use Multicolor detectors	Nonuniformity over large area High cost in growth and processing Surface and bulk instabilities
		III–V (InGaAs, InAs, InSb, InAsSb)	Good-quality material and dopants Advanced technology Possible monolithic integration	Heteroepitaxy with large lattice mismatch Long wavelength cutoff limited to 7 μm (at 77 K)
	Extrinsic (Si:Ga, Si:As, Ge:Cu, Ge:Hg)		Very long wavelength operation Relatively simple technology	High thermal generation Extremely low-temperature operation
	Free carriers (PtSi, Pt$_2$Si, IrSi)		Low cost, high yields Large and close packed 2D arrays	Low quantum efficiency Low temperature operation
	Quantum wells	Type I (GaAs/AlGaAs, InGaAs/AlGaAs)	Mature material growth Good uniformity over large area Multicolor detectors	High thermal generation Complicated design and growth
		Type II (InAs/InGaSb, InAs/InAsSb)	Low Auger recombination rate Easy wavelength control Multicolor detectors	Complicated design and growth Sensitive to the interfaces
	Quantum dots	InAs/GaAs, InGaAs/InGaP, Ge/Si	Normal incidence of light Low thermal generation	Complicated design and growth

power, and low-cost coolers. Thus, it is highly desirable to eliminate or reduce the cooling requirements in an IR system.

Can thermal detectors replace photon detectors in most applications where uncooled operation is essential? The answer is simply—no! Thermal detectors seem to be less suitable for the next generation of IR thermal imaging systems, which are moving toward faster frame rates and multispectral operation; they are completely useless for many imaging and nonimaging applications that require nano- and subnano-second responses.

Photodetectors make it possible to achieve both high sensitivity and fast response (see, e.g., Refs. 2, 3). They are devices that detect radiation by direct interaction of photons with electrons in the detector material. This interaction generates free-charge carriers in intrinsic or extrinsic detectors, or delivers the necessary energy to charge carriers confined within a potential well to overcome the barrier.

A number of concepts to improve performance of photodetectors operating at near room temperatures have been proposed.[4–6] Recent results show that, in principle, the ultimate limits of sensitivity, even for wavelengths exceeding 10 µm, can be achieved without the need for cryogenic cooling.[7, 8]

1.2 Detector Figures of Merit

It is difficult to measure the performance characteristics of IR detectors because of the large number of experimental variables involved. A variety of environmental, electrical, and radiometric parameters must be taken into account and carefully controlled. With the advent of large 2D detector arrays, detector testing has become even more complex and demanding. Numerous texts and journals cover this issue, including: *Infrared System Engineering* by R. D. Hudson;[9] *The Infrared Handbook* edited by W. L. Wolfe and G. J. Zissis;[10] *The Infrared and Electro-Optical Systems Handbook* edited by J. S. Accetta and D. L. Shumaker;[11] and *Fundamentals of Infrared Detector Operation and Testing* by J. D. Vincent.[12] In this volume we have restricted our consideration to detectors whose output consists of an electrical signal that is proportional to the radiant signal power.

The measured data described in this text are sufficient to characterize a detector. However, to provide ease of comparison between detectors, certain figures of merit, computed from the measured data, will be defined in this section.

1.2.1 Responsivity

The responsivity of an IR detector is defined as the ratio of the root mean square (rms) value of the fundamental component of the electrical output signal of the detector to the rms value of the fundamental component of the input radiation power. The units of responsivity are volts per watt (V/W) or amperes per watt (amp/W).

The voltage (or analogous current) spectral responsivity is given by

$$R_v(\lambda, f) = \frac{V_s}{\Phi_e(\lambda)\Delta\lambda},$$

(1.1)

where V_s is the signal voltage due to Φ_e, and $\Phi_e(\lambda)$ is the spectral radiant incident power (in W/m).

An alternative to the above monochromatic quality, the blackbody responsivity, is defined by the equation

$$R_v(T, f) = \frac{V_s}{\Phi_{ebb}} = \frac{V_s}{\displaystyle\int_0^\infty \Phi_e(\lambda)d\lambda}, \tag{1.2}$$

where the incident radiant power is the integral over all wavelengths of the spectral density of power distribution $\Phi_e(\lambda)$ from a blackbody. The responsivity is usually a function of the bias voltage V_b, the operating electrical frequency f, and the wavelength λ.

1.2.2 Noise equivalent power

The noise equivalent power (NEP) is the incident power on the detector generating a signal output equal to the rms noise output. Stated another way, the NEP is the signal level that produces a signal-to-noise ratio (SNR) of 1. It can be written in terms of responsivity as

$$\text{NEP} = \frac{V_n}{R_v} = \frac{I_n}{R_i}. \tag{1.3}$$

The unit of NEP is the watt.

The NEP is also quoted for a fixed reference bandwidth, which is often assumed to be 1 Hz. This "NEP per unit bandwidth" has a unit of watts per square root hertz ($\text{W/Hz}^{1/2}$).

1.2.3 Detectivity

The detectivity D is the reciprocal of NEP:

$$D = \frac{1}{\text{NEP}}. \tag{1.4}$$

It was found by Jones[13] that for many detectors, the NEP is proportional to the square root of the detector signal that is proportional to the detector area, A_d. This means that both NEP and detectivity are functions of the electrical bandwidth and detector area, so a normalized detectivity D^* (or "D-star") suggested by Jones[13, 14] is defined as

$$D^* = D(A_d \Delta f)^{1/2} = \frac{(A_d \Delta f)^{1/2}}{\text{NEP}}. \tag{1.5}$$

The importance of D^* is that this figure of merit permits a comparison of detectors that have different areas. Either a spectral or blackbody D^* can be defined in terms of the corresponding type of NEP.

Useful equivalent expressions to Eq. (1.5) include:

$$D^* = \frac{(A_d \Delta f)^{1/2}}{V_n} R_v = \frac{(A_d \Delta f)^{1/2}}{I_n} R_i = \frac{(A_d \Delta f)^{1/2}}{\Phi_e} (\text{SNR}), \qquad (1.6)$$

where D^* is defined as the rms SNR in a 1-Hz bandwidth per unit rms incident radiant power per square root of detector area. D^* is expressed by the unit $\text{cm}\,\text{Hz}^{1/2}\,\text{W}^{-1}$, which also has been called a "Jones."

Spectral detectivity curves for a number of commercially available IR detectors are shown in Fig. 1.4. Interest has centered mainly on the wavelengths of the two atmospheric windows 3–5 µm (middle wavelength infrared or MWIR) and 8–14 µm (the LWIR region). Atmospheric transmission is the highest in the MWIR and LWIR bands, and the emissivity maximum of the objects at $T \approx 300$ K is at the

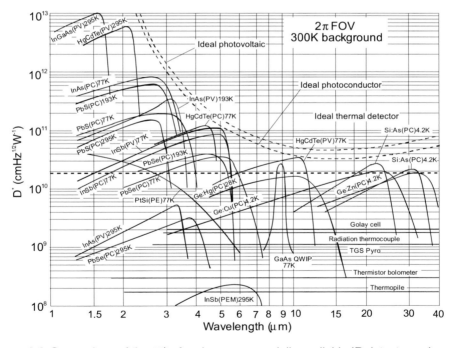

Figure 1.4 Comparison of the D^* of various commercially available IR detectors when operated at the indicated temperature. Chopping frequency is 1000 Hz for all detectors except the thermopile (10 Hz), thermocouple (10 Hz), thermistor bolometer (10 Hz), Golay cell (10 Hz), and pyroelectric detector (10 Hz). Each detector is assumed to view a hemispherical surrounding at a temperature of 300 K. Theoretical curves for the background-limited D^* (dashed lines) for the ideal photovoltaic and photoconductive detectors and for thermal detectors are also shown. Key: PC = photoconductive detector, PV = photovoltaic detector, PE = photoemissive detector, and PEM = photoelectromagnetic detector.

wavelength $\lambda \approx 10$ μm. However, in recent years space applications have increased the interest in longer wavelengths.

The blackbody $D^*(T, f)$ may be found from spectral detectivity:

$$D^*(T, f) = \frac{\int_0^\infty D^*(\lambda, f)\Phi_e(T, \lambda)d\lambda}{\int_0^\infty \Phi_e(T, \lambda)d\lambda} = \frac{\int_0^\infty D^*(\lambda, f)E_e(T, \lambda)d\lambda}{\int_0^\infty E_e(T, \lambda)d\lambda}, \quad (1.7)$$

where $\Phi_e(T, \lambda) = E_e(T, \lambda)A_d$ is the incident blackbody radiant flux (in W), and $E_e(T, \lambda)$ is the blackbody irradiance (in W/m^2).

The ultimate performance of IR detectors is reached when the detector and amplifier noise are low compared to the photon noise. The photon noise is fundamental in the sense that it arises not from any imperfection in the detector or its associated electronics, but rather from the detection process itself as a result of the discrete nature of the radiation field.

1.3 Detectivity Requirements for Thermal Imagers

For focal plane arrays (FPAs), the relevant figure of merit is the noise equivalent temperature difference (NEDT), the temperature change of a scene required to produce a signal equal to the rms noise. The configuration of a basic thermal imager system is shown in Fig. 1.5.

It can be shown that the NETD of an IR imager is

$$\text{NETD} = \frac{4F_\#^2 \Delta f^{1/2}}{A_d^{1/2} M^*}, \quad (1.8)$$

where $F_\#$ is the optics f-number, Δf is the frequency band, A_d is the detector area, and M^* is the thermal figure of merit, which is dependent on the detectivity of

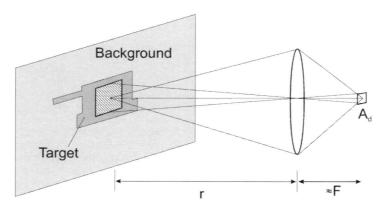

Figure 1.5 Thermal imager system configuration.

detector $D^*(\lambda)$ and spectral blackbody exitance $R(\lambda, T)$, described by the Planck law between two wavelengths λ_1 and λ_2 as

$$M^* = \int_{\lambda_1}^{\lambda_2} D^*(\lambda) \frac{\partial R(\lambda, T)}{\partial T} d\lambda, \tag{1.9}$$

$$R(\lambda, T) = \frac{2\pi hc^2}{\lambda^5} \left[\exp\left(\frac{hc}{\lambda kT}\right) - 1 \right]^{-1}, \tag{1.10}$$

where $c = 2.998 \times 10^8$ m/s is the light velocity, $h = 6.6256 \times 10^{-34}$ J s is the Planck's constant, and $k = 1.38054 \times 10^{-23}$ J/K is the Boltzmann's constant.[14, 15]
For nonselective detectors,

$$M^* = D^* \int_{\lambda_1}^{\lambda_2} \frac{\partial R(\lambda, T)}{\partial T} d\lambda. \tag{1.11}$$

Infrared imaging systems with staring arrays demand less bandwidth when compared to scanning systems. Reducing pixel size is important for lightweight imagers, but a higher D^* would be necessary to compensate for NETD deterioration with smaller pixel size [see Eqs. (1.8) and (1.11)].

Table 1.2 shows the $\int_{\lambda_1}^{\lambda_{co}} (\lambda/\lambda_{co})[\partial R(\lambda, T)/\partial T] d\lambda$ and D^* required for NETD = 0.1 K as a function of spectral range for a staring IR imager with relatively fast optics ($F_\# = 1$), operating with a standard frame rate (50 frames/sec).

The detectivity required for a thermal detector operating over the entire spectral range is 0.65×10^8 cm Hz$^{1/2}$/W. Reducing the spectral range to the 8–14 µm atmospheric window increases this detectivity by a factor of ≈ 2.3. Operation in the 3–5 µm atmospheric window would require a significant increase of detectivity—by more than one order of magnitude. This makes thermal detectors not very useful for operation outside the LWIR atmospheric window.

Table 1.2 The $\int_{\lambda_1}^{\lambda_{co}} (\lambda/\lambda_{co})[\partial R(\lambda, T)/\partial T] d\lambda$ and detectivity required for NETD = 0.1 K ($F_\# = 1$, $T = 300$ K, detector size 50×50 µm).

Spectral range (µm)	$\int_{\lambda_1}^{\lambda_{co}} \frac{\lambda}{\lambda_{co}} \frac{\partial R(\lambda, T)}{\partial T} d\lambda$ (W/m^{2}K)	D^* (cm Hz$^{1/2}$/W)
0–infinity	6.12	0.65×10^8
0–14	4.25	0.94×10^8
8–14	2.63	1.52×10^8
3–5	0.186	2.15×10^9

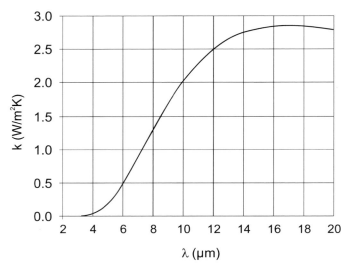

Figure 1.6 Dependence of $k = \int_0^{\lambda_{co}} (\lambda/\lambda_{co})[\partial R(\lambda, T)/\partial T]d\lambda$ on wavelength; $T = 300$ K.

Consider an ideal photon counter detector with cutoff wavelength of λ_{co}. In this case,

$$M^* = D^*(\lambda_{\max}) \int_0^{\lambda_{co}} \frac{\lambda}{\lambda_{co}} \frac{\partial R(\lambda, T)}{\partial T} d\lambda. \qquad (1.12)$$

Figure 1.6 shows $\int_0^{\lambda_{co}} (\lambda/\lambda_{co})[\partial R(\lambda, T)/\partial T]d\lambda$ as a function of wavelength. This function initially increases with wavelength, achieving its maximum just above the 8–14 μm spectral range band.

Let's consider NETD = 0.1 K as a minimal requirement for thermal imagers. As Fig. 1.7 shows, detectivity measures of 1.9×10^8 cm Hz$^{1/2}$/W, 2.3×10^8 cm Hz$^{1/2}$/W, and 2×10^9 cm Hz$^{1/2}$/W are necessary for an NETD of 0.1 K for, respectively, the 10 μm, 9 μm, and 5 μm cutoff wavelength photon counter detectors. Very good thermal imaging will require an NETD by a factor 3 to 10 less. Can this be achieved at near room temperature using photodetectors? The answer to this question is given in Chapter 4.

1.4 Cooling of IR Detectors

The method of cooling varies according to the operating temperature and the system's logistical requirements.[16, 17] Various types of cooling systems have been developed including dewars with cryogenic liquids or solids, Joule-Thompson open cycle, Stirling closed cycle, and thermoelectric coolers (see Fig. 1.8). These systems are discussed briefly below.

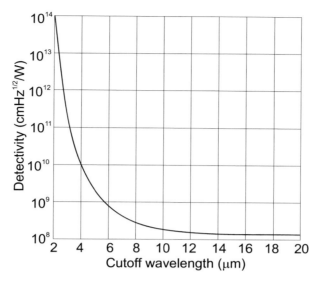

Figure 1.7 Dependence of detectivity on cutoff wavelength for a photon counter detector with thermal resolution of 0.1 K.

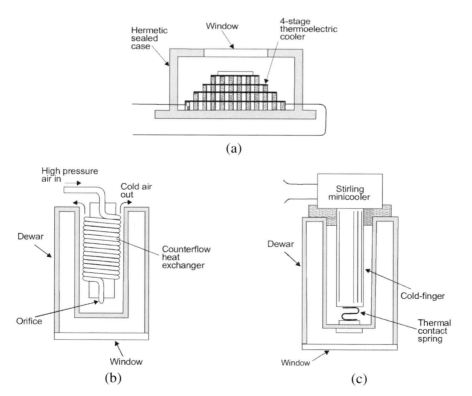

Figure 1.8 Three ways of cooling IR detectors: (a) four-stage thermoelectric cooler (Peltier effect), (b) Joule-Thompson cooler, and (c) Stirling-cycle engine.

1.4.1 Cryogenic dewars

Most 8–14-μm detectors operate at approximately 77 K and can be cooled by liquid nitrogen. Cryogenic liquid pour-filled dewars are frequently used for detector cooling in laboratories. They are rather bulky and need to be refilled with liquid nitrogen every few hours. For many applications, especially in the field LN_2, pour-filled dewars are impractical, so many manufacturers are turning to alternative coolers that do not require cryogenic liquids or solids.

1.4.2 Joule-Thompson coolers

The design of Joule-Thompson coolers is based on the fact that as a high-pressure gas expands upon leaving a throttle valve, it cools and liquefies. The coolers require a high-pressure gas supply from bottles and compressors. The gas used must be purified to remove water vapor and carbon dioxide that could freeze and block the throttle valve. Specially designed Joule-Thompson coolers using argon are suitable for ultrafast cool-down.

1.4.3 Stirling cycle coolers

Ever since the late 1970s, military systems have overcome the problem of LN_2 operation by utilizing Stirling closed-cycle refrigerators to generate the cryogenic temperatures necessary for critical IR detector components. These refrigerators were designed to produce operating temperatures of 77 K directly from DC power. Early versions were large and expensive, and suffered from microphonic and electromagnetic interference (EMI) noise problems. Today, smaller and more efficient integral and split Stirling coolers have been developed and refined.

The use of cooling engines, in particular those employing the Stirling cycle, has increased recently due to their efficiency, reliability, and cost reduction. Stirling engines require several minutes of cool-down time; the working fluid is helium. Both Joule-Thompson and engine-cooled detectors are housed in precision-bore dewars into which the cooling device is inserted (see Fig. 1.8). The detector, mounted in the vacuum space at the end of the inner wall of the dewar and surrounded by a cooled radiation shield compatible with the convergence angle of the optical system, looks out through an IR window. In some dewars, the electrical leads to detector elements are embedded in the inner wall of the dewar to protect them from damage due to vibration.

Typical Stirling cooler parameters are:

- For integral: mass—0.3 kg, heat load—0.15 W, input power—3.5 W, shelf life—5 years, mean time before failure (MTBF) = 2000 hr, size with detector—9 × 9 cm, 80 K achievable in 4 minutes, significant microphonics. After 2000 to 3000 hours of operation, this cooler will require factory service to maintain performance. Because the dewar and cooler is one integrated unit, the entire unit must be serviced together.

- For split: input power $\approx$ 20 W, 1.5 W heat load, fewer microphonics, 8-year MTFB. The detector is mounted on the dewar bore, and the cold finger of the cooler is thermally connected to the dewar by a bellows. A fan is necessary to dissipate the heat. The cooler can be easily removed from the detector/dewar for replacement.

Closed-cycle coolers offer the ultimate detector performance without the need for bulk liquid nitrogen. However, the high initial cost, large input power, and limited operating life are not suited for many applications.

1.4.4 Peltier coolers

Thermoelectric (TE) cooling of detectors is simpler and less costly than closed-cycle cooling. In the case of Peltier coolers, detectors are usually mounted in a hermetic encapsulation with a base designed to make good contact with a heat sink. TE coolers can achieve temperatures to $\approx$ 200° K, have $\approx$ 20-year operating life and low input power ($<$1 W for a 2-stage device and $<$3 W for a 3-stage device), and are small and rugged. Peltier coolers can be used to stabilize the temperature at the required level. There are hopes for much better performance of TE coolers based on superlattice materials.[18] With the TE figure of merit, ZT, almost three times larger than that in Bi_2Te_3, temperatures of $<$200 K would be achievable with simple 2-stage TE coolers based on the new materials.

References

1. P. W. Kruse and D. D. Skatrud (Eds.), *Semiconductors and Semimetals*, Vol. 47, Academic Press, San Diego (1997).
2. A. Rogalski (Ed.), *Infrared Photon Detectors*, Vol. PM20, SPIE Press, Bellingham, WA (1995).
3. A. Rogalski, K. Adamiec, and J. Rutkowski, *Narrow-Gap Semiconductor Photodiodes*, Vol. PM77, SPIE Press, Bellingham, WA (2000).
4. C. T. Elliott and N. T. Gordon, "Infrared Detectors," in *Handbook on Semiconductors*, Vol. 4, C. Hilsum (Ed.), North-Holland, Amsterdam, 841–936 (1993).
5. J. Piotrowski, W. Galus, and M. Grudzień, "Near room-temperature IR photodetectors," *Infrared Phys.* **31**, 1–48 (1991).
6. J. Piotrowski, "$Hg_{1-x}Cd_xTe$ Infrared Photodetectors," in *Infrared Photon Detectors*, Vol. PM20, 391–494, SPIE Press, Bellingham, WA (1995).
7. C. T. Elliott, N. T. Gordon, and A. M. White, "Towards background-limited, room-temperature, infrared photon detectors in the 3–13 μm wavelength range," *Appl. Phys. Lett.* **74**, 2881–2883 (1999).
8. M. A. Kinch, "Fundamental physics of infrared detector materials," *J. Electron. Mater.* **29**, 809–817 (2000).
9. R. D. Hudson, *Infrared System Engineering*, Wiley, New York (1969).

10. W. I. Wolfe and G. J. Zissis (Eds.), *The Infrared Handbook*, Office of Naval Research, Washington, D.C. (1985).

11. W. D. Rogatto (Ed.), *The Infrared and Electro-Optical Systems Handbook*, Infrared Information Analysis Center, Ann Arbor, MI, and SPIE Press, Bellingham, WA (1993).

12. J. D. Vincent, *Fundamentals of Infrared Detector Operation and Testing*, Wiley, New York (1990).

13. R. C. Jones, "Performance of detectors for visible and infrared radiation," in *Advances in Electronics*, Vol. 5, L. Morton (Ed.), Academic Press, New York, 27–30 (1952).

14. E. L. Dereniak and G. D. Boreman, *Infrared Detectors and Systems*, Wiley, New York (1996).

15. A. Rogalski, *Infrared Detectors*, Gordon and Breach, Amsterdam (2000).

16. J. L. Miller, *Principles of Infrared Technology*, Van Nostrand Reinhold, New York (1994).

17. P. T. Blotter and J. C. Batty, "Thermal and mechanical design of cryogenic cooling systems," in *The Infrared and Electro-Optical Systems Handbook*, Vol. 3, 343–433, W. D. Rogatto (Ed.), Infrared Information Analysis Center, Ann Arbor, MI, and SPIE Press, Bellingham, WA (1993).

18. R. J. Radtke, H. Ehrenreich, and C. H. Grein, "Multilayer thermoelectric refrigeration in $Hg_{1-x}Cd_xTe$ superlattices," *J. Appl. Phys.* **86**, 3195–3198 (1999).

Chapter 2

Fundamental Performance Limitations of Infrared Photodetectors

This chapter discusses the fundamental limitations to IR photodetector performance imposed by the statistical nature of the generation and recombination processes in the semiconductor material, and radiometric considerations. We will try to establish the ultimate theoretical sensitivity limit that can be expected for a detector operating at a given temperature. The model presented here is applicable to any of the photodetector classes mentioned in Chapter 1. We will also attempt to describe how to improve the performance of the device by taking into account fundamental physics. The nonfundamental limitations, such as Shockley-Read processes, contact phenomena, bias power dissipation, high electric fields, dislocations, fabrication problems, current gain, and quantum efficiency peculiarities, will be addressed later in this chapter.

2.1 Infrared Photon Detector Classifications

As discussed in Chapter 1, IR photodetectors are usually classed as one of five types based on the type of optical transitions involved in the process of absorption of radiation[1]:

1. Intrinsic
2. Extrinsic
3. Photoemissive
4. Quantum wells
5. Quantum dots

Recently this standard classification was reconsidered by Kinch,[2] as discussed briefly below.

Photon detectors can be divided into two broad classes, namely majority and minority carrier devices. The material systems used are:

1. Direct bandgap semiconductors—minority carriers

    • Binary alloys: InSb, InAs
    • Ternary alloys: HgCdTe, InGaAs
    • Type II, III superlattices: InAs/GaInSb, HgTe/CdTe

2. Extrinsic semiconductors—majority carriers

    • Si:As, Si:Ga, Si:Sb
    • Ge:Hg, Ge:Ga

3. Type I superlattices—majority carriers

    • GaAs/AlGaAs quantum well infrared photodetectors (QWIPs)

4. Silicon Schottky barriers—majority carriers

    • PtSi, IrSi

5. High-temperature superconductors—minority carriers

All of these material systems have been serious players in the IR marketplace with the exception of high-temperature superconductors.

2.2 Theoretical Model

Let us consider a generalized model of a photodetector, which by its optical area A_o is coupled to a beam of IR radiation.[3–6] The detector is a slab of homogeneous semiconductor with actual "electrical" area, A_e, and thickness t (see Fig. 2.1). Usually, the optical and electrical areas of the device are the same or similar. However,

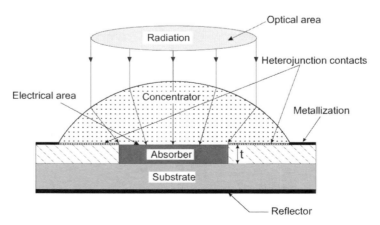

Figure 2.1 Model of a photodetector.

the use of some kind of optical concentrator can increase the A_o/A_e ratio by a large factor.

The current responsivity of the photodetector is determined by the quantum efficiency, η, and by the photoelectric gain, g. The quantum efficiency describes how well the detector is coupled to the impinging radiation. It is defined here as the number of electron-hole pairs generated per incident photon in an intrinsic detector, the number of generated free unipolar charge carriers in an extrinsic detector, or the number of charge carriers with energy sufficient to cross the potential barrier in a photoemissive detector. The photoelectric gain is the number of carriers passing contacts per one generated pair in an intrinsic detector, or the number of charge carriers in other types of detectors. This value shows how well the generated charge carriers are used to generate current response of a photodetector. Both values are assumed here as constant over the volume of the device.

The spectral current responsivity is equal to

$$R_i = \frac{\lambda \eta}{hc} qg, \qquad (2.1)$$

where λ is the wavelength, h is the Planck's constant, c is the velocity of light, q is the electron charge, and g is the photoelectric current gain. The current that flows through the contacts of the device is noisy due to the statistical nature of the generation and recombination processes—fluctuation of optical generation, thermal generation, and radiative and nonradiative recombination rates. Assuming that the current gain for the photocurrent and the noise current are the same, the noise current is

$$I_n^2 = 2q^2 g^2 (g_{op} + g_{th} + r) \Delta f, \qquad (2.2)$$

where g_{op} is the optical generation rate, g_{th} is the thermal generation rate, r is the resulting recombination rate, and Δf is the frequency band.

It should be noted that the effects of a fluctuating recombination frequently can be avoided by arranging for the recombination process to take place in a region of the device where it has little effect due to a low photoelectric gain—for example, at the contacts in sweep-out photoconductors, at the backside surface of a photo-electromagnetic detector, or in the neutral regions of the diodes. The generation processes with their associated fluctuations, however, cannot be avoided by any means.[7, 8]

Detectivity, D^*, is the main parameter to characterize normalized signal-to-noise performance of detectors, and can be defined as

$$D^* = \frac{R_i (A_o \Delta f)^{1/2}}{I_n}. \qquad (2.3)$$

2.2.1 Optical generation noise

Optical generation noise is photon noise due to fluctuation of the incident flux. The optical generation of the charge carriers may result from three different sources:

- signal radiation generation
- background radiation generation
- thermal self-radiation of the detector itself at a finite temperature.

2.2.1.1 Noise due to optical signal

The optical signal generation rate (photons/s) is

$$g_{op} = \Phi_s A_o \eta, \tag{2.4}$$

where Φ_s is the signal photon flux density.

If recombination does not contribute to the noise,

$$I_n^2 = 2\Phi_s A_o \eta q^2 g^2 \Delta f \tag{2.5}$$

and

$$D^* = \frac{\eta^{1/2}}{2^{1/2}\Phi_s^{1/2}hc}. \tag{2.6}$$

This is the ideal situation, when the noise of the detector is determined entirely by the noise of the signal photons. Usually, the noise due to optical signal flux is small compared to the contributions from background radiation or thermal generation-recombination processes. An exception is heterodyne detection, when the noise due to the powerful local oscillator radiation may dominate.

2.2.1.2 Noise due to background radiation

Background radiation frequently is the main source of noise in a detector. Assuming no contribution due to recombination,

$$I_n^2 = 2\Phi_B A_o \eta q^2 g^2 \Delta f, \tag{2.7}$$

where Φ_B is the background photon flux density. Therefore,

$$D^*_{\text{BLIP}} = \frac{\lambda \eta^{1/2}}{hc\Phi_B^{1/2}}. \tag{2.8}$$

Once background-limited performance is reached, quantum efficiency, η, is the only detector parameter that can influence a detector's performance.

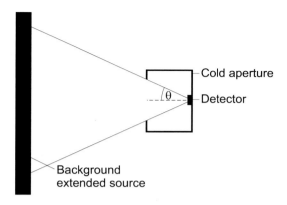

Figure 2.2 Use of cold aperture to reduce a detector's field of view.

The background photon flux density is controlled by background temperature, its emissivity, and the detector's field of view (FOV), which can be limited by a cold shield. When circular cold aperture is used to limit a detector's FOV, as shown in Fig. 2.2, then

$$\Phi_B = \pi L_B \sin^2 \theta, \qquad (2.9)$$

where L_B is the background radiance.

Consider a detector that is a photon counter with cutoff wavelength λ_{co}. For blackbody background radiation,

$$L_B = 2c \int_0^{\lambda_{co}} \lambda^{-4} \left[\exp\left(\frac{hc}{\lambda k T} \right) - 1 \right]^{-1} d\lambda \qquad (2.10)$$

and

$$\Phi_B = 2\pi c \sin^2 \theta \int_0^{\lambda_{co}} \lambda^{-4} \left[\exp\left(\frac{hc}{\lambda k T} \right) - 1 \right]^{-1} d\lambda. \qquad (2.11)$$

Figure 2.3 shows the peak spectral D^*_{BLIP} of a photon counter versus the cutoff wavelength plot calculated for 300 K background radiation and hemispherical FOV ($\theta = 90$ deg). The minimum D^*_{BLIP} (300 K) occurs at 14 μm and is equal to 4.6×10^{10} cm Hz$^{1/2}$/W. For some photodetectors that operate at near-equilibrium conditions, such as non-sweep-out photoconductors, the recombination rate is equal to the generation rate. For these detectors the contribution of recombination to the noise will reduce D^*_{BLIP} by a factor of $2^{1/2}$. Note that D^*_{BLIP} does not depend on area and the A_o/A_e ratio. As a consequence, the background-limited and signal-limited performances cannot be improved by making A_o/A_e large.

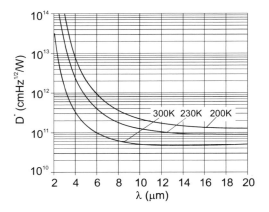

Figure 2.3 Calculated peak spectral detectivities of a photon counter limited by the hemispherical FOV background radiation as a function of the peak wavelength and temperature.

2.2.1.3 Internal radiative generation

In contrast to the signal and background related processes, optical generation is connected with the detector itself and may be of importance for detectors operating at near room temperatures. The related ultimate performance is usually calculated, assuming blackbody radiation and taking into account the reduced speed of light and wavelength due to a greater than 1 refractive index of the detector material and full absorption of photons with energy larger than the bandgap.[9] The carrier generation rate per unity area is

$$g_a = 8\pi c n^2 \int_0^\infty \frac{d\lambda}{\lambda^4 (e^{hc/\lambda T} - 1)}, \tag{2.12}$$

where n is the refractive index. Note that the generation rate is a factor of $4n^2$ larger compared to the 180-deg FOV background generation for $\eta = 1$ [compare to Eq. (2.11)]. Therefore, the resulting detectivity

$$D^* = \frac{\lambda \eta A_o}{hc (2g_a A_e)^{1/2}} \tag{2.13}$$

will be a factor of $2n$ lower when compared to BLIP detectivity for a background at detector temperature ($A_e = A_o$). The internal thermal radiation limited D^* can be improved by making A_o/A_e large, in contrast to D^*_{BLIP}.

When Humpreys re-examined the existing theories of radiative recombination and internal optical generation,[10, 11] he indicated that most photons emitted as a result of radiative recombination are immediately reabsorbed inside the detector, generating charge carriers. Due to reabsorption, the radiative lifetime is highly extended, which means that the internal optical generation-recombination processes could be practically noiseless in optimized devices. Therefore, the ultimate limit of performance is set by the signal or background photon noise.

2.2.2 Thermal generation and recombination noise

Infrared photodetectors operating at near room temperature and low-temperature devices operated at low background irradiances are generally limited by thermal generation and recombination mechanisms rather than by photon noise. For effective absorption of IR radiation in a semiconductor, we must use material with a low energy of optical transitions compared to the energy of photons to be detected—for example, semiconductors with a narrower band gap. A direct consequence of this fact is that at near room temperatures, the thermal energy of charge carriers, kT, becomes comparable to the transition energy. This enables thermal transitions, making the thermal generation rate very high. As a result, the long-wavelength detector is very noisy when operated at near room temperature.

For uniform volume generation and recombination rates G and R (in $m^{-6} s^{-1}$), the noise current is

$$I_n^2 = 2(G + R)A_e t \Delta f q^2 g^2; \tag{2.14}$$

therefore,

$$D^* = \frac{\lambda}{2^{1/2}hc(G + R)^{1/2}} \left(\frac{A_o}{A_e}\right)^{1/2} \frac{\eta}{t^{1/2}}. \tag{2.15}$$

At equilibrium, the generation and recombination rates are equal. In this case,

$$D^* = \frac{\lambda\eta}{2hc(Gt)^{1/2}} \left(\frac{A_o}{A_e}\right)^{1/2}. \tag{2.16}$$

2.3 Optimum Thickness of a Detector

For a given wavelength and operating temperature, the highest performance can be obtained by maximizing $\eta/[(G + R)t]^{1/2}$. This is the condition for the highest ratio of the quantum efficiency to the square root of the sum of the sheet thermal generation and recombination rates. This means that high quantum efficiency must be obtained with a thin device.

In the following calculations, we will assume $A_e = A_o$, perpendicular incidence of radiation, and negligible front and backside reflection coefficients. In this case,

$$\eta = 1 - e^{-\alpha t}, \tag{2.17}$$

where α is the absorption coefficient. Then

$$D^* = \frac{\lambda}{2^{1/2}hc} \left(\frac{\alpha}{G + R}\right)^{1/2} F(\alpha t), \tag{2.18}$$

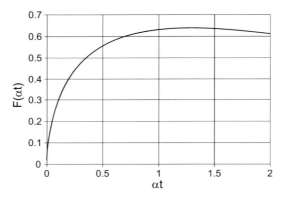

Figure 2.4 The plot of function $F(\alpha t) = (1 - e^{-\alpha t})(\alpha t)^{-1/2}$.

where

$$F(\alpha t) = \frac{1 - e^{-\alpha t}}{(\alpha t)^{1/2}}. \tag{2.19}$$

The plot of $F(\alpha)$ is shown in Fig. 2.4. Function $F(\alpha t)$ achieves its maximum 0.638 for $t = 1.26/\alpha$. In this case $\eta = 0.716$, and the highest detectivity is

$$D^* = 0.45 \frac{\lambda}{hc} \left(\frac{\alpha}{G + R} \right)^{1/2}. \tag{2.20}$$

The detectivity can also be increased by a factor of $2^{1/2}$ for a double pass of radiation. This can be achieved with the use of a backside reflector. Simple calculation shows that the optimum thickness in this case is half that of the single pass case, while the quantum efficiency remains equal to 0.716.

At equilibrium, the generation and recombination rates are equal. Therefore,

$$D^* = \frac{\lambda}{2hc} \eta (Gt)^{-1/2}. \tag{2.21}$$

The effects of a fluctuating recombination can frequently be avoided by arranging for the recombination process to take place in a region of the device where it has little effect due to low photoelectric gain: for example, at the contacts in sweep-out photoconductors or in the neutral regions of diodes. In this case, the noise can be reduced and the detectivity increased by a factor of $2^{1/2}$. However, the generation process, with its associated fluctuation, cannot be avoided by any means. If the recombination process is uncorrelated with the generation process, it contributes the detector noise

$$D^* = \frac{\lambda}{2^{1/2}hc} \eta (Gt)^{-1/2}. \tag{2.22}$$

2.4 Detector Material Figure of Merit

To summarize the discussion above, the detectivity of an optimized IR photodetector of any type can be expressed as

$$D^* = 0.31 \frac{\lambda}{hc} k \left(\frac{\alpha}{G} \right)^{1/2},$$ (2.23)

where $1 \leq k \leq 2$ and is dependent on the contribution of recombination and backside reflection, as shown in Table 2.1.

As we can see, the ratio of the absorption coefficient to the thermal generation rates is the main figure of merit of any IR detector materials. This figure of merit, proposed for the first time by Piotrowski, can be utilized to predict ultimate performance of any IR detector and to select possible material candidates for use as detectors.[4, 12]

The α/G ratio versus temperature for different types of tunable materials with a hypothetical energy gap equal to 0.25 eV ($\lambda = 5$ μm) and 0.124 eV ($\lambda = 10$ μm) is shown in Fig. 2.5. Procedures used to calculate α/G for different material systems are given in Ref. 13. It is apparent that HgCdTe is by far the most efficient detector

Table 2.1 Dependence of the factor k in Eq. (2.23), optimum thickness, and quantum efficiency on contribution of recombination and presence of backside reflection.

Backside reflection	Contribution of recombination	Optimum thickness	Quantum efficiency	k
0	$R = G$	$1.26/\alpha$	0.716	1
1	$R = G$	$0.63/\alpha$	0.716	$2^{1/2}$
0	No	$1.26/\alpha$	0.716	$2^{1/2}$
1	No	$0.63/\alpha$	0.716	2

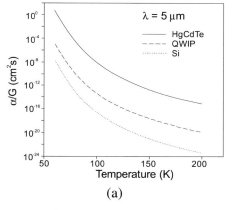
(a)

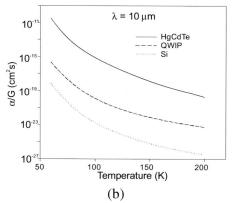
(b)

Figure 2.5 α/G ratio versus temperature for (a) MWIR − $\lambda = 5$ μm, and (b) LWIR − $\lambda = 10$-μm photon detectors. (Reprinted from Ref. 14 with permission from the IOP.)

of IR radiation. One may also notice that a QWIP is a better material than extrinsic silicon.

It should be noted that the importance of the thermal generation rate as a material figure of merit was recognized for the first time by Long.[15] It was used in many papers by English workers[16, 17] related to high operating temperature (HOT) detectors. More recently Kinch[12] introduced the thermal generation rate within $1/\alpha$ depth per unit of area as the figure of merit, which is actually the inverse α/G figure of merit originally proposed by Piotrowski.[12]

The calculation of the figure of merit requires a determination of the absorption coefficient and the thermal generation rate, taking into account various processes of both a fundamental and a less fundamental nature.

2.5 Reducing Device Volume to Enhance Performance

One possible way to improve the performance of IR photodetectors is to reduce the total amount of thermal generation within the active element of the detector by reducing the detector volume, which is the product of its thickness and physical area. This must be done:

- Without sacrificing quantum efficiency,
- By preserving the required optical area, and
- By keeping the acceptance angle of the detector large enough to intercept radiation from the IR system's main optics.

Interestingly, the performance of thermal detectors is related in a similar way to its physical area when the thermal conductance to the surrounding area is constant.

2.5.1 Enhancing absorption

Enhancing absorption makes it possible to reduce a detector's thickness without losing quantum efficiency. The simplest way to enhance absorption is the use of a retroreflector to double pass in the IR radiation, as considered in Sec. 2.4. However, there are more efficient ways.

The quantum efficiency of thin devices can be significantly enhanced by using interference phenomena to set up a resonant cavity within the photodetector.[18–23] Various optical resonator structures are shown in Fig. 2.6. In the simplest method, interference occurs between the waves reflected at the rear, highly reflective surface, and at the front surface of the semiconductor. The thickness of the semiconductor is selected to set up the standing waves in the structure with peaks at the front and nodes at the back surface. The quantum efficiency oscillates with the thickness of the structure, and the thickness of the peaks corresponds to an odd multiple of $\lambda/4n$, where n is the refractive index of the semiconductor. The gain in quantum efficiency increases with n. Higher gain can be obtained in the structures shown in Fig. 2.6(b). With the use of interference effects, a strong and high

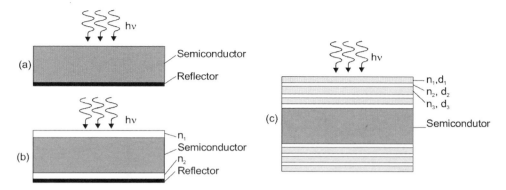

Figure 2.6 Schematic structure of interference-enhanced quantum efficiency in a photodetector: (a) the simplest structure, (b) a structure immersed between two dielectric layers and supplied with a backside reflector, and (c) a structure immersed between two photonic crystals.

nonuniform absorption can be achieved even for long-wavelength radiation with a low absorption coefficient. For example, a 0.9 quantum efficiency is obtainable in the optimized Au/ZnS/HgCdTe/ZnS structures with a HgCdTe thickness as low as ≈ 1.7 μm, which normally will have a quantum efficiency less than 0.2.[19] Even greater improvement is possible in structures with multiple dielectric layers, such as that shown in Fig. 2.6(c).

Enhanced absorption due to interference effects seems to be particularly important for devices whose operation is dependent on the gradient of minority carriers, such as Dember and electromagnetic effect detectors.[3, 19, 24–26] An optimized resonant cavity structure with carefully selected thicknesses and reflection coefficients of all layers exhibits a detectivity enhanced by a factor of ≈ 2–4 at peak wavelength compared to optimized conventional devices.

It should be noted that the interference effects strongly modify the spectral response of a device, and the gain due to the optical cavity can be achieved only in narrow spectral regions. This may be an important limitation for applications that require a wide spectral band sensitivity. In practice, IR systems usually operate in a spectral band (e.g., atmospheric windows), and the use of resonant cavities may yield significant gains. Another limitation comes from the fact that efficient optical resonance occurs only for perpendicular incidence and is less effective for oblique incidence. This limits the use of devices with fast optics, and especially devices with optical immersion.[27, 28]

2.5.2 Increasing the apparent "optical" area of a detector compared to its physical area

Another possible means of improving the performance of an IR photodetector is to increase the apparent "optical" size of the detector in comparison with its actual physical size using a suitable concentrator that compresses the impinging IR radiation. This must be achieved without reducing the acceptance angle, or at

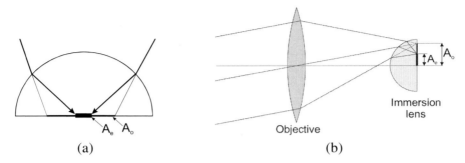

Figure 2.7 (a) Principle of optical immersion, and (b) ray tracing for an optical system with a combination of an objective lens and an immersion lens.

least with limited reduction of the angles required by the fast optics of IR systems. Various types of suitable optical concentrators can be used, including optical cones, conical fibers, and other types of reflective, diffractive, and refractive optical concentrators.[3, 17, 29–31]

An efficient way to achieve an effective concentration of radiation is to adhere the photodetector to a hemispherical or a hyperhemispherical immersion lens[32] with the detector located at the center of curvature of the lens. The lens produces an image of the detector, and no spherical or coma aberration exists (aplanatic imaging). Due to this immersion, the apparent linear size of the detector increases by a factor of n. The image is produced at the detector plane. The principle of operation of a hemispherical immersion lens is shown in Fig. 2.7.

The use of a hemispheric immersion lens in combination with an objective lens of an imaging optical system is shown in Fig. 2.7(b). The immersion lens plays the role of a field lens, which increases the FOV of the optical system.

The limit to the compression is determined by the Lagrange invariant ($A\Omega$ product) and the sin condition for an aplanatic system.[31] In air, the physical and the apparent size of the detector are related by the equation

$$n^2 A_e \sin \theta' = A_o \sin \theta, \qquad (2.24)$$

where n is the lens refractive index; A_e and A_o are the physical and apparent sizes of the detector, respectively; and θ and θ' are the marginal ray angles before refraction at the lens and at the image, respectively. Therefore,

$$\frac{A_o}{A_e} = n^2 \frac{\sin \theta'}{\sin \theta}. \qquad (2.25)$$

For the hemispherical lens, the marginal ray angles are 90 deg. Therefore, the area gain is n^2. Larger gain can be obtained for a hyperhemisphere used as an aplanatic lens (Fig. 2.8). This results in an apparent increase in the linear detector size by a factor of n^2. In this case, the image plane is shifted.

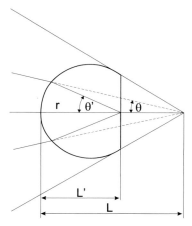

Figure 2.8 Hyperimmersion lens, where θ and θ' are the ray angles before refraction at the lens and at the image, respectively, and $\sin\theta' = n\sin\theta$.

The size of the detector and of its image in relation to the front of the hyperimmersion lens are

$$L' = \left(1 + \frac{1}{n}\right)r \tag{2.26}$$

and

$$L = (n + 1)r. \tag{2.27}$$

The condition for minimum $F_\#$ is

$$\sin\theta_{\max} = \frac{1}{n} = (4F^2 - 1)^{-1/2}. \tag{2.28}$$

Hence,

$$F_{\min} = \frac{(n^2 + 1)^{1/2}}{2}. \tag{2.29}$$

Figure 2.9 shows the dependence of $F_{\min}$ on the refraction index.

Several factors must be taken into account for practical realization of optically immersed detectors. The practical use of immersion technology has been limited due to problems with mechanical matching of the detector and lens materials, and to severe transmission and reflection losses. Another limitation was due to the limited acceptance angle of the devices as a result of the total reflection at the lens-glue interface.

Germanium ($n = 4$) is the most frequently used material for immersion lenses. Arsenic-doped amorphous selenium or Mylar has been used for the adhesive.[33]

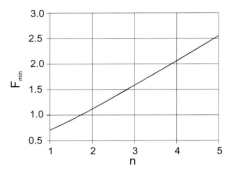

Figure 2.9 Dependence of the minimum $F_\#$ of hyperhemispherical immersion lenses on the refraction index.

Silicon is another important immersion lens material. The problem of matching the detector material to the immersion lens material has been solved by the use of monolithic technology developed at VIGO System S.A.[22, 34] The technology is based on the epitaxy of HgCdZnTe on transparent and high refraction index CdZnTe substrates. The HgCdZnTe serves as the sensitive element, while the immersion lens is formed directly in the substrate. This technology also has been used for InGaAs and InAsSb devices on GaAs substrates.[22, 23, 34]

As Table 2.2 shows, monolithic optical immersion results in the significant improvement of a detector's parameters. The gain factors achieved with hyperhemispherical immersion are substantially higher compared to those for hemispherical immersion, although the hyperhemispherical immersion may restrict the acceptance angle of the detector and require more severe manufacturing tolerances. These restrictions depend on the refraction coefficient of the lens. For CdZnTe they are not so severe as for germanium lenses, however, and have no practical impor-

Table 2.2 Influence of optical immersion on properties of photodetectors with a CdZnTe lens. The numbers show the relative change of a given value compared to a nonimmersed detector of the same optical size.

Value	Hemisphere		Hyperhemisphere		Type of detector
Linear size	n	(≈ 2.7)	n^2	(≈ 7)	Any
Area	n^2	(≈ 7)	n^4	(≈ 50)	Any
Voltage responsivity	n	(≈ 2.7)	n^2	(≈ 7)	PC*, PEM
Voltage responsivity	n^2	(≈ 7)	n^4	(≈ 50)	PV**, Dember
Detectivity	n	(≈ 2.7)	n^2	(≈ 7)	Any
Bias power	n^2	(≈ 7)	n	(≈ 50)	PC, PV
Capacitance	n^2	(≈ 7)	n^4	(≈ 50)	PV
Acceptance angle (deg)	180		42		Any
Thickness/radius	1		1,3		Any
Tolerances	Not severe		Stringent		Any

*PC is a photoconductive detector.
**PV is a photovoltaic detector.

tance in many cases. For example, the minimum f-number of the main optical system is limited to about 1.4 for the CdZnTe immersion lens.

An alternative approach is the use of a Winston cone. QinetiQ has developed a micromachining technique involving dry etching to fabricate the cone concentrators for detector and luminescent devices.[35–37]

References

1. A. Rogalski, *Infrared Detectors*, Gordon and Breach, Amsterdam (2000).
2. M. A. Kinch, "Fundamental physics of infrared detector materials," *J. Electron. Mater.* **29**, 809–817 (2000).
3. J. Piotrowski, "$Hg_{1-x}Cd_xTe$ infrared photodetectors," in *Infrared Photon Detectors*, Vol. PM20, 391–494, SPIE Press, Bellingham, WA (1995).
4. J. Piotrowski and W. Gawron, "Ultimate performance of infrared photodetectors and figure of merit of detector material," *Infrared Phys. Technol.* **38**, 63–68 (1997).
5. J. Piotrowski and A. Rogalski, "New generation of infrared photodetectors," *Sensors and Actuators A* **67**, 146–152 (1998).
6. J. Piotrowski, "Uncooled operation of IR photodetectors," *Opto-Electr. Rev.* **12**, 111–122 (2004).
7. T. Ashley and C. T. Elliott, "Non-equilibrium mode of operation for infrared detection," *Electron. Lett.* **21**, 451–452 (1985).
8. T. Ashley, T. C. Elliott, and A. M. White, "Non-equilibrium devices for infrared detection," *Proc. SPIE* **572**, 123–132 (1985).
9. S. Jensen, "Temperature limitations to infrared detectors," *Proc. SPIE* **1308**, 284–292 (1990).
10. R. G. Humpreys, "Radiative lifetime in semiconductors for infrared detectors," *Infrared Phys.* **23**, 171–175 (1983).
11. R. G. Humpreys, "Radiative lifetime in semiconductors for infrared detectors," *Infrared Phys.* **26**, 337–342 (1986).
12. J. Piotrowski, "$Hg_{1-x}Cd_xTe$: material for the present and future generation of infrared sensors," *MST News Poland* No. **1**, 4–5 (March 1997).
13. A. Rogalski, "Quantum well photoconductors in infrared detectors technology," *J. Appl. Phys.* **93**, 4355–4391 (2003).
14. A. Rogalski, "HgCdTe infrared detector material: History, status, and outlook," *Rep. Prog. Phys.* **68**, 2267–2336 (2005).
15. D. Long, "Photovoltaic and photoconductive infrared detectors," in *Optical and Infrared Detectors*, R. J. Keyes (Ed.), Springer-Verlag, Berlin, 101–147 (1977).
16. C. T. Elliott and N. T. Gordon, "Infrared detectors," in *Handbook on Semiconductors*, Vol. 4, C. Hilsum (Ed.), North-Holland, Amsterdam, 841–936 (1993).
17. C. T. Elliott, "Photoconductive and non-equilibrium devices in HgCdTe and related alloys," in *Infrared Detectors and Emitters: Materials and Devices*, 279–

312, P. Capper and C.T. Elliott (Eds.), Kluwer Academic Publishers, Boston (2001).

18. D. L. Spears, "10.6 μm photomixer arrays at 195 K," in *Proc. IRIS Active Systems*, 331–349 (1984).

19. J. Piotrowski, W. Galus, and M. Grudzień, "Near-room-temperature IR photodetectors," *Infrared Phys.* **31**, 1–48 (1990).

20. M. S. Ünlü and S. Strite, "Resonant cavity enhanced photonic devices," *J. Appl. Phys.* **78**, 607–639 (1995).

21. Z. Djuric, Z. Jaksic, D. Randjelowic, T. Dankocic, W. Ehrfeld, and A. Schmidt, "Enhancement of radiative lifetime in semiconductors using photonic crystals," *Infrared Phys. Technol.* **40**, 25–32 (1999).

22. J. Kaniewski, Z. Orman, J. Piotrowski, M. Sioma, L. Ornoch, and M. Romanis, "Epitaxial InAs detectors optically immersed to GaAs microlenses," *Proc. SPIE* **4369**, 721–730 (2001).

23. J. Piotrowski, H. Mucha, Z. Orman, J. Pawluczyk, J. Ratajczak, and J. Kaniewski, "Refractive GaAs microlenses monolithically integrated with InGaAs and HgCdTe photodetectors," *Proc. SPIE* **5074**, 918–925 (2003).

24. D. Genzow, A. Józwikowska, K. Józwikowski, T. Niedziela, and J. Piotrowski, "Photoelectromagnetic effect (PEM) in graded gap structures (numerical approach)," in *Proc. Conf. on Physics of Semiconductor Compounds* **6**, Ossolineum, Wroclaw, Poland, 316–320 (1983).

25. D. Genzow, K. Józwikowski, M. Nowak, and J. Piotrowski, "Interference photoelectromagnetic effect in $Hg_{1-x}Cd_x$Te graded gap structures," *Infrared Phys.* **24**, 21–24 (1984).

26. M. Nowak, "Photoelectromagnetic effect in semiconductors and its applications," *Prog. Quant. Electron.* **11**, 205–346 (1987).

27. M. Sioma and J. Piotrowski, "Modelling and optimization of high temperature detectors of long wavelength IR radiation with optical resonant cavity," *Opto-Electr. Rev.* **12**, 157–160 (2004).

28. J. Kaniewski, J. Muszalski, J. Pawluczyk, and J. Piotrowski, "Resonant cavity enhanced InGaAs photodiodes for high speed detection of 1.55 μm infrared radiation," *Proc. SPIE* **5783**, 47–56 (2005).

29. J. E. Slavek and H. H. Randal, "Optical immersion of HgCdTe photoconductive detectors," *Infrared Phys.* **15**, 339–340 (1975).

30. M. Grudzień and J. Piotrowski, "Monolithic optically immersed HgCdTe IR detectors," *Infrared Phys.* **29**, 251–253 (1989).

31. J. Piotrowski, M. Grudzień, Z. Nowak, et al., "Uncooled photovoltaic $Hg_{1-x}Cd_x$Te LWIR detectors," *Proc. SPIE* **4130**, 175–184 (2000).

32. R. C. Jones, "Immersed radiation detectors," *App. Opt.* **1**, 607–613 (1962).

33. E. L. Dereniak and G. D. Boreman, *Infrared Detectors and Systems*, Wiley, New York (1996).

34. T. T. Piotrowski, A. Piotrowska, E. Kamińska, et al., "Design and fabrication of GaSb/InGaAsSb/AlGaAsSb mid-infrared photodetectors," *Opto-Electron. Rev.* **9**, 188–194 (2001).

35. T. Ashley, D. T. Dutton, C. T. Elliott, N. T. Gordon, and T. J. Phillips, "Optical concentrators for light emitting diodes," *Proc. SPIE* **3289**, 43–50 (1998).

36. M. K. Haigh, G. R. Nash, N. T. Gordon, et al., "Progress in negative luminescent $Hg_{1-x}Cd_xTe$ diode arrays," *Proc. SPIE* **5783**, 376–383 (2005).

37. G. J. Bowen, I. D. Blenkinsop, R. Catchpole, et al., "HOTEYE: A novel thermal camera using higher operating temperature infrared detectors," *Proc. SPIE* **5783**, 392–400 (2005).

Chapter 3

Materials Used for Intrinsic Photodetectors

The operation of the intrinsic IR photodetector is based on band-to-band transitions caused by absorption of the photons in a narrow-gap semiconductor. Among the many types of IR photodetectors, intrinsic photodetectors offer the best performance at high temperatures.

The wide body of information now available concerning different methods of crystal growth and physical properties of materials used for IR photodetectors makes it difficult to review all aspects in detail. As a result, only selected topics are reviewed in this chapter. More information can be found in many comprehensive reviews and monographs (see, for example, Refs. 1–8).

The properties of narrow-gap semiconductors that are used as the material systems for IR detectors result from the direct energy bandgap structure: a high density of states in the valence and conduction bands, which results in strong absorption of IR radiation and a relatively low rate of thermal generation. Table 3.1 compares important parameters of narrow-gap semiconductors used in IR detector fabrication.

Table 3.1 Some physical properties of narrow-gap semiconductors.

Material	E_g (eV)		n_i (cm^{-3})		ε	μ_e (10^4 cm^2/V s)		μ_h (10^4 cm^2/V s)	
	77 K	300 K	77 K	300 K		77 K	300 K	77 K	300 K
InAs	0.414	0.359	6.5×10^3	9.3×10^{14}	14.5	8	3	0.07	0.02
InSb	0.228	0.18	2.6×10^9	1.9×10^{16}	17.9	100	8	1	0.08
In$_{0.53}$Ga$_{0.47}$As	0.66	0.75		5.4×10^{11}	14.6	7	1.38		0.05
PbS	0.31	0.42	3×10^7	1.0×10^{15}	172	1.5	0.05	1.5	0.06
PbSe	0.17	0.28	6×10^{11}	2.0×10^{16}	227	3	0.10	3	0.10
PbTe	0.22	0.31	1.5×10^{10}	1.5×10^{16}	428	3	0.17	2	0.08
Pb$_{1-x}$Sn$_x$Te	0.1	0.1	3.0×10^{13}	2.0×10^{16}	400	3	0.12	2	0.08
Hg$_{1-x}$Cd$_x$Te	0.1	0.1	3.2×10^{13}	2.3×10^{16}	18.0	20	1	0.044	0.01
Hg$_{1-x}$Cd$_x$Te	0.25	0.25	7.2×10^8	2.3×10^{15}	16.7	8	0.6	0.044	0.01

3.1 Semiconductors for Intrinsic Photodetectors

Table 3.2 shows the most important material systems used for intrinsic photodetectors. These systems include fixed bandgap binary alloys, tunable bandgap ternary semiconductors, and bandgap engineered superlattice materials. The binary compounds can be used for applications that require optimum performance at the spectral range corresponding to the bandgap of the material. However, the availability of binary compounds is very limited, and no compound is known to operate in the LWIR spectral range. Therefore, the use of tunable bandgap ternary and quaternary materials is necessary for many applications.

There are no clear indications that some materials among the binary alloys and tunable bandgap semiconductors are better in terms of fundamental figure of merit α/G. It is expected that some artificial narrow gap semiconductors based on type II and type III superlattices may exhibit an improved figure of merit due to inherent Auger suppression caused by the specific band structure.[1]

$Hg_{1-x}Cd_xTe$ is undeniably the champion among a large variety of material systems. Three key features secure its position:

- A tailorable energy bandgap over the 1–30-μm range,
- Large optical coefficients that enable a high quantum efficiency, and
- Favorable inherent recombination mechanisms that lead to a high operating temperature.

These properties are a direct consequence of the energy band structure of this zinc-blende semiconductor. Moreover, the specific advantages of HgCdTe are its ability to obtain both low and high carrier concentrations, the high mobility of its electrons, and its low dielectric constant.

The extreme flexibility of the HgCdTe material system for IR applications enables the fabrication of an optimized detector of any type, for detection in any region of the IR spectrum, including dual and multicolor devices. This flexibility is due to the tunability of the bandgap that covers the whole IR spectral range and the tunability of n- and p-type doping over a very wide range (10^{14}–10^{19} cm^{-3}). But perhaps the decisive advantage of $Hg_{1-x}Cd_xTe$ is the independence of its lattice parameters on composition. Among variable bandgap semiconductor alloys,

Table 3.2 The most important material systems used for intrinsic photodetectors.

Material system	Most important	Others
Binary alloys	InSb, InAs	PbS, PbSe
Tunable bandgap semiconductors	Hg-based	$Hg_{1-x}Cd_xTe$, $Hg_{1-x}Zn_xTe$, $Hg_{1-x}Mn_xTe$
	Lead salts	$Pb_{1-x}Sn_xTe$, $Pb_{1-x}Sn_xSe$
	InSb-based	InAsSb, InNSb, InBiTe, InTlSb
Type II superlattices	InAs/GaInSb	
Type III superlattices	HgTe/CdTe	

$Hg_{1-x}Cd_xTe$ is the only material that covers the wide spectral band and has nearly the same lattice parameters. The difference between CdTe ($E_g = 1.5$ eV) and $Hg_{0.8}Cd_{0.2}Te$ ($E_g = 0.1$ eV) is $\approx 0.2\%$. Replacing a small fraction of Cd with Zn, or Te with Se can completely compensate the residual lattice mismatch. This is especially important because recent IR photodetector development has been dominated by complex bandgap heterostructures that are necessary to improve the performance of multicolor devices.

The growth of bulk crystals and epitaxial layers of $Hg_{1-x}Cd_xTe$ and closely related alloys has been extensively reported in numerous original and review papers. Perhaps the most important sources are the publications from annual (since 1981) workshop-style technical meetings devoted exclusively to the physics and chemistry of $Hg_{1-x}Cd_xTe$ and related semiconductor and IR materials. This information was published initially in *Journal of Vacuum Science and Technology* **21**(1) (1982), **A1**(3) (1983), **A3**(1) (1985), **A4**(4) (1986), **A5**(5) (1987), **A6**(4) (1988), **A7**(2) (1989), **B9**(3) (1991), and **B10**(4) (1992); in the American Institute of Physics Conference Proceeding no. 235 (1991); and in *Journal of Electronic Materials* **22**(8) (1993), **24**(5) (1995), **24**(9) (1995), **25**(8) (1996), **26**(6) (1997), **27**(6) (1998), **28**(6) (1999), **29**(6) (2000), **30**(6) (2001), **31**(7) (2002), **32**(7) (2003), **33**(6) (2004), **34**(6) (2005), and **35**(6) (2006). Excellent early review papers on characterization methods were published in *Proceedings of the International Workshop on Mercury Cadmium Characterization* (1992) and *Semiconductor Science and Technology* **8**(6S) (1993). Other important sources are *Materials Research Society Symposia Proceedings* **90** (1987), **161** (1990), **216** (1991), **299** (1993), **450** (1996), **484** (1998), and **607** (2000); *IEE Conferences on Advanced Infrared Detectors and Systems*, Conference Publication nos. **228** (1981), **263** (1986), and **321** (1990) (IEE London); *Narrow-Gap Semiconductor Conferences* (1990, 1991, and 1993); and numerous *Proceedings of SPIE*. Most of the important data on $Hg_{1-x}Cd_xTe$ properties can be found in reviews edited by Capper[2, 3] and Rogalski.[1, 4]

3.2 $Hg_{1-x}Cd_xTe$ Ternary Alloys

The performance of IR photodetectors is determined by the following fundamental properties of the semiconductor used: the bandgap, the intrinsic concentration, the mobilities of electrons and holes, the absorption coefficient, and the thermal generation and recombination rates. Table 3.3 contains a list of important material parameters.

3.2.1 Band structure and electrical properties

3.2.1.1 Band structure

At present the band structure of $Hg_{1-x}Cd_xTe$ is well established. The electrical and optical properties of $Hg_{1-x}Cd_xTe$ are determined by the energy gap structure in the vicinity of the Γ-point of the Brillouin zone, in essentially the same way as

Table 3.3 Physical properties of $Hg_{1-x}Cd_xTe$. Data were taken from Refs. 5–7.

Property	T (K)	$x = 0$	$x = 0.2$	$x = 1$
Lattice constant a (nm)	300	0.6463	0.64645	0.6482
Thermal expansion coeff. α (10^{-6}/K)	300	4	4.3	5.5
Thermal conductivity C (W/cm K)	300	0.04		0.07
Density γ (g/cm^3)	300	8.076	7.63	5.85
Melting point T_m (K)		943	940 (sol.)	1365
			1050 (liq.)	
Bandgap E_g (eV)	300	−0.15	0.165	1.505
	77	−0.25	0.09	1.60
	4.2	−0.30	0.064	1.605
Effective masses: m_e^*/m	77	0.029	0.005	0.096
m_h^*/m		0.35–0.7	0.3–0.7	0.66
Mobilities (cm^2/V s): μ_e	77		2.5×10^5	4×10^4
μ_h			7×10^2	3.8×10^4
Intrinsic carrier concentration n_i (cm^{-3})	300		3×10^{16}	
	77		8×10^{13}	
Static dielectric constant ε_s	300	20.5	18.5	10.5
High-frequency dielectric constant ε_∞	300	15.0	13.0	7.2

InSb.[1, 5, 7, 8] The energy gap of this compound ranges from −0.300 eV for semi-metallic HgTe, goes through zero at approximately $x = 0.15$, and opens up to 1.648 eV for CdTe.

Figure 3.1 plots the energy bandgap $E_g(x, T)$ for $Hg_{1-x}Cd_xTe$ versus the alloy composition parameter x at temperature 77 K and 300 K. Also plotted is the cutoff wavelength $\lambda_c(x, T)$, defined as that wavelength at which the response has dropped to 50% of its peak value.

A number of expressions approximating $E_g(x, T)$ are available at present.[1–3] Here we use the expression given by Seiler et al.[9] that gives the best fit to the experimental data:

$$E_g = -0.302 + 1.93x - 0.81x^2 + 0.832x^3 + 5.32 \times 10^{-4}(1 - 2x)\left(\frac{-1822 + T^3}{255.2 + T^2}\right),$$
$$(3.1)$$

where E_g is in eV and T is in K.

3.2.1.2 Intrinsic concentration

The expressions for intrinsic concentrations have been derived from theoretical calculations, taking into account nonparabolicity, using the k · p method.[10] The intrinsic concentration as a function of composition and temperature can be approximated as

$$n_i = \left(A_g + B_g x + C_g T + D_g xT + F_g x^2 + G_g T^2\right) \times 10^{14} E_g^{3/4} T^{3/2} \exp\left(\frac{-E_g}{2kT}\right),$$
$$(3.2)$$

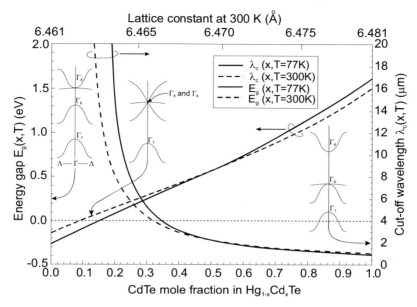

Figure 3.1 The bandgap structure of $Hg_{1-x}Cd_x$Te near the Γ-point for three different values of the forbidden energy gap. The energy bandgap is defined as the difference between the Γ_6 and Γ_8 band extreme at $\Gamma = 0$.

where n_i is in cm^{-3}, $A_g = 5.24256$, $B_g = -3.5729$, $C_g = -4.74019 \times 10^{-4}$, $D_g = 1.25942 \times 10^{-2}$, $F_g = 5.77046$, and $G_g = 4.24123 \times 10^{-6}$. The electron m_e^* and light hole m_{lh}^* effective masses in the narrow-gap Hg compounds are close and they can be established according to the Kane band model. Here we use the following expression:

$$\frac{m}{m_e^*} = 1 + 2F + \frac{E_p}{3}\left(\frac{2}{E_g} + \frac{1}{E_g + \Delta}\right), \tag{3.3}$$

where $E_p = 19$ eV, $\Delta = 1$ eV, and $F = -0.8$.[11] The effective mass of heavy hole m_{hh}^* is high; the measured values range between 0.3–0.7 m.[12] The value of $m_{hh}^* = 0.55$ m is frequently used to model IR detectors.

3.2.1.3 Mobilities

Due to small effective masses, the electron mobilities in $Hg_{1-x}Cd_x$Te are remarkably high while heavy-hole mobilities are two orders of magnitude lower. A number of scattering mechanisms influence the electron mobility including ionized impurities, alloy disorder, electron-electron and hole-hole events, acoustic phonon modes, and polar phonon modes.[13] Nonpolar optical phonon scattering is significant in p-type and semimetallic n-type materials. Although calculations of electron mobilities are generally in good agreement with the experiment, there is still no general theoretical understanding of hole mobilities in $Hg_{1-x}Cd_x$Te.

The electron mobility in $Hg_{1-x}Cd_xTe$ ($T > 50$ K) can be approximated as:

$$\mu_e = \frac{9 \times 10^4 s}{T^{2r}} \qquad \text{(in m}^2\text{/V s)},[14] \qquad (3.4)$$

with $s = 0.2^{7.5}/x^{7.5}$ and $r = 0.2^{0.6}/x^{0.6}$. For $T = 300$ K,

$$\mu_e = (8.754x - 1.044)^{-1} \qquad \text{(in m}^2\text{/V s).}[15] \qquad (3.5)$$

The ratio of electron to hole mobility is

$$b = \frac{\mu_e}{\mu_h} = \frac{610}{(E_g T)^{1/2}} \qquad (E_g \text{ in V and } T \text{ in K).}[16] \qquad (3.6)$$

For modeling IR photoconductors, the hole mobility is usually calculated assuming that the electron-to-hole mobility ratio b is constant and equal to 100.

3.2.2 Defects and impurities

Native defect properties and impurity incorporation still constitute a field of intensive research. Various aspects of defects in bulk crystals and epilayers, such as electrical activity, segregation, ionization energies, diffusivity, and carrier lifetimes have been summarized in excellent reviews.[1, 17–24]

3.2.2.1 Native defects

The defect structure of undoped and doped $Hg_{1-x}Cd_xTe$ can be explained with the quasichemical approach.[19, 25–30] The dominant native defect in $Hg_{1-x}Cd_xTe$ is a double ionizable acceptor associated with metal lattice vacancies. Some direct measurements show much larger vacancy concentrations than those that follow from Hall measurements, indicating that most vacancies are neutral.[31]

In contrast to numerous early findings, it seems to now be established that the native donor defect concentration is negligible. As-grown undoped and pure $Hg_{1-x}Cd_xTe$, including that grown in Hg-rich liquid phase epitaxy (LPE), always exhibits p-type conductivity with the hole concentration depending on composition, growth temperature, and Hg pressure during growth reflecting correspondence to the concentration of vacancies.

The equilibrium concentration of vacancies and Hg pressures over Te-saturated $Hg_{1-x}Cd_xTe$ are

$$C_V\left[\text{cm}^{-3}\right] = \left(5.08 \times 10^{27} + 1.1 \times 10^{28}x\right)$$
$$\times P_{Hg}^{-1} \exp\left[\frac{-(1.29 + 1.36x - 1.8x^2 + 1.375x^3)\text{ eV}}{kT}\right], \quad (3.7)$$

$$P_{Hg}[\text{atm}] = 1.32 \times 10^5 \exp\left(-\frac{0.635\text{ eV}}{kT}\right).[27] \qquad (3.8)$$

The Hg pressure over Hg saturated $Hg_{1-x}Cd_xTe$ is close to saturated Hg pressure:

$$P_{Hg}[atm] = (5.0 \times 10^6 + 5.0 \times 10^6 x) \times \exp\left(\frac{-0.99 + 0.25x}{kT} \text{ eV}\right). \quad (3.9)$$

Figure 3.2 shows the hole concentration as a function of the partial Hg pressure showing $1/p_{Hg}$ dependence of the native acceptor concentration, which is in agreement with the predictions of the quasichemical approach for narrow-gap $Hg_{1-x}Cd_xTe$. Annealing in Hg vapors reduces the hole concentration by filling the vacancies. Low-temperature (<300°C) annealing in Hg vapors reveals the background impurity level, causing the p-to-n conversion in some crystals. Samples with higher residual donor concentration turn n-type at higher temperatures and show higher electron concentration. Unexpected effects may arise from Te precipitates.[32] Hg diffusing into material dissolves precipitates and drives the major impurities ahead of Hg, leaving the core p-type. On further annealing, these impurities may redistribute throughout the slice, turning the whole sample p-type. A variety of effects may cause unexpected n-type behavior contamination, surface layers formed during cool-down, strain, dislocations, twins, grain boundaries, substrate orientation, oxidation, and perhaps other parameters.

Native defects play a dominant role in the diffusion behavior.[33] Vacancies have very high diffusivities even at low temperatures. For example, to form a junction a few micrometers deep in 10^{16} cm^{-3} material requires only about 15 minutes at a temperature of 150–200°C. This corresponds to diffusion constants of the order of 10^{-10} cm^{-2}/s. The presence of dislocations can enhance vacation mobility even further, while the presence of Te precipitates may retard the motion of Hg into lattice.

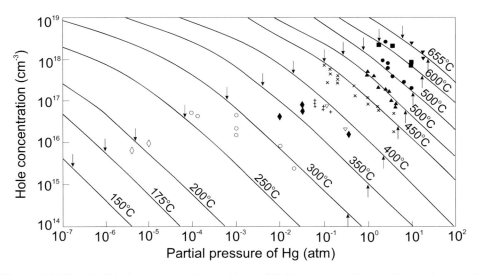

Figure 3.2 The 77 K hole concentration in $Hg_{1-x}Cd_xTe$ calculated according to the quasichemical approach as a function of the partial pressure of Hg. Arrows define the material existence region. (Reprinted from Ref. 18 with permission from AIP.)

3.2.2.2 Dopants

The electrical behavior of dopants has been extensively reviewed in two review papers.[1, 22] Donor behavior is expected for elements from group IIIB on the metal lattice site, and group VIIB elements in the Te site. Indium is most frequently used as a well-controlled dopant for n-type doping due to its high solubility and moderately high diffusion. The experimental data can be explained, assuming that at low ($<10^{18}$ cm^{-3}) concentration In incorporates as a single ionizable donor occupying a metal lattice site. At high In concentration, In incorporates as a neutral complex corresponding to In_2Te_3. The bulk materials are typically doped by direct addition to melts. Indium is frequently introduced during epitaxy and by diffusion; it has been used for many years as a contact material for the n-type photoconductors and the n-type side of photodiodes.

Among the group VIIB elements, only I that was occupying Te sites proved to be a well behaved donor with concentrations in the 10^{15}–10^{18} cm^{-3} range.[21, 34–36] The electron concentration was found to increase with Hg pressure. Acceptor behavior is expected of elements in the I group (Ag, Cu, and Au) substituting for metal lattice sites, and of elements in the V group (P, As, Sb, Bi) substituting for Te sites.

Ag, Cu, and Au are shallow single acceptors.[17, 22] They are very fast diffusers that limit the applications for devices. Significant diffusion of Ag and especially Cu occurs at room temperature.[37] Hole concentrations have been obtained roughly equal to Cu concentration: up to 10^{19} cm^{-3}. But the behavior of Au is more complex. Au seems to be not very useful as a controllable acceptor, though it has proven to be useful for contacts.

The amphoteric behavior of the VB group elements (P, As, Sb) has been established.[21, 35] They are acceptors substituting for Te sites and donors at metal sites; therefore, metal-rich conditions are necessary to introduce dopants at Te sites. Arsenic proved to be the most successful p-type dopant to date for formation of stable junctions.[38–41] The main advantages are very low diffusity, stability in lattice, low activation energy, and the possibility of controlling concentration over a wide (10^{15}–10^{18} cm^{-3}) range. Intensive efforts are currently underway to reduce the high temperature ($\approx400°C$) and high Hg pressures required to activate arsenic as an acceptor.

3.2.3 Growth of bulk crystals and epilayers

High-quality semiconductor material is essential to the production of high-performance and affordable IR photodetectors. The material must have a low defect density, large wafer size, uniformity, and reproducibility of intrinsic and extrinsic properties. To achieve these characteristics, $Hg_{1-x}Cd_xTe$ materials evolved from high-temperature, melt-grown, bulk crystals to low-temperature, liquid and vapor phase epitaxy. However, the cost and availability of large-area and high-quality $Hg_{1-x}Cd_xTe$ are still the main considerations for producing affordable devices.

3.2.3.1 Phase diagrams

A solid understanding of phase diagrams is essential to properly design the growth process. The phase diagrams and their implications for $Hg_{1-x}Cd_xTe$ crystal growth have been discussed extensively.[6, 24, 42, 43] The ternary Hg-Cd-Te phase diagrams have been established both theoretically and experimentally throughout the Gibbs triangle.[44–46] Brice[45] summarized the works of more than 100 researchers on the Hg-Cd-Te phase diagram as a numerical description, which is convenient for the design of growth processes; for a recent review, see Ref. 22.

The generalized associated solution model has proven to be successful to explain experimental data and to predict the phase diagram of the entire Hg-Cd-Te system. It was assumed that the liquid phase was a mixture of Hg, Te, Cd, HgTe, and CdTe [see Fig. 3.3(b)]. The gas phase over the material contained Hg, Cd atoms, and Te_2 molecules. The composition of the solid material can be described by a generalized formula $(Hg_{1-x}Cd_x)_{1-y}Te_y$. The familiar $Hg_{1-x}Cd_xTe$ formula corresponds to the pseudobinary CdTe and HgTe alloy ($y = 0.5$) with complete mutual solubilities. At present it is believed that the sphalerite pseudobinary phase region in $Hg_{1-x}Cd_xTe$ is extended in Te-rich material with a width of the order of 1%. The width narrows at lower temperatures. The consequence of such a form of diagram is a tendency for Te precipitation. An excess of Te is due to vacancies in the metal sublattice, which results in p-type conductivity of pure materials. Low-temperature annealing at 200–300 K reduces the native defect (predominantly acceptor) concentration and reveals an uncontrolled (predominantly donor) impurity background. Weak Hg-Te bonding results in low activation energy for defect formation and Hg migration in the matrix. This can cause bulk and surface instabilities.

Most of the problems with crystal growth are due to the marked difference between the solidus and liquidus curves (see Figs. 3.3 and 3.4) resulting in the segregation of binaries during crystallization from melts. The segregation coefficient

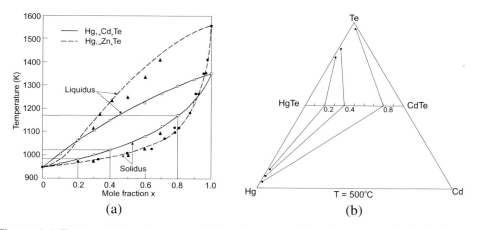

Figure 3.3 Ternary phase diagrams: (a) liquidus and solidus lines in the HgTe-CdTe and HgTe-ZnTe pseudobinary systems; (b) the Gibbs diagram of the Hg-Cd-Te system.

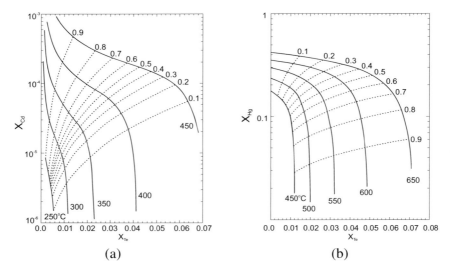

Figure 3.4 Liquidus isotherms for (a) Hg-rich, and (b) Te-rich growth. (Reprinted from Ref. 44 with permission from Elsevier.)

for growth from melts depends on Hg pressure. Serious problems also arise from high Hg pressures over pseudobinary and Hg-rich melts. Figure 3.4 shows the low-temperature liquidus and solid-solution isoconcentration lines in the Hg-rich and Te-rich corners. Figure 3.4(a) shows that, e.g., at 450°C, a solid solution containing 0.90 mole fraction CdTe is in equilibrium with a liquid containing 7×10^{-4} atomic fraction Cd and 0.014 atomic fraction Te. One sees that almost pure CdTe(s) crystallizes from very Hg-rich liquids.

3.2.3.2 Bulk crystals

Bulk crystals were used to fabricate IR detectors for first-generation imaging and seeking systems, mostly for n-type single-element photoconductors, linear arrays, SPRITE detectors, and photodiodes. Good uniformity, excellent electrical purity (less than 10^{14} cm^{-3}), high mobilities, and minority carrier lifetimes are standard features for commercially available materials. The cost of the device-quality material is still high, ranging between $300–1000 for a 1.5-cm-diameter wafer, depending on composition, doping, and extent of characterization.

Several bulk crystal growth techniques have been used for high-volume production of the material. As the phase diagrams show, a strong segregation occurs during crystallization from melts unless the freezing proceeds very quickly. Thus, the conventional melt growth techniques, such as the standard Bridgman method, produce crystals with large longitudinal and axial variations of composition. These crystals also are not very useful to fabricate device-quality materials. Instead, an improved Bridgman-Stockbarger growth process has been favored at Mullard.[47]

The quench-anneal method (also called the modified Bridgman method and solid state recrystallization) has been arguably the best-known production-type growth method of uniform and high-quality crystals.[6, 48] In this method the charge of a required composition is synthesized, melted, and quenched. Then the fine dendritic mass obtained in such process is annealed below the liquidus temperature for a few weeks to recrystallize and homogenize the crystals. Various improvements of the process have been proposed including temperature gradient annealing and slow cooling to prevent Te precipitation. The material usually requires low-temperature annealing to adjust the concentration of native defects. The crystals also can be uniformly doped by introducing dopants to the charge.

The quench-anneal method has some drawbacks. Since no segregation occurs, all impurities present in the charge are frozen in the crystal, and high-purity starting elements are required. The maximum diameter of the ingots is limited to approximately 1.5 cm because the cooling rate of the large-diameter charges is too slow to suppress the segregation. The crystals also contain low grain boundaries.

The relatively homogeneous and highly perfect crystals can be grown in systems in which depletion of the liquid in the wide-bandgap binary material during growth is compensated by continuous replenishment from solid "slush" particles suspended in the upper part of the growth melt.[6]

Large and homogeneous crystals with diameters up to ≈ 5 cm can be grown using zone melting. The growth temperature can be substantially reduced when growing crystals from Te-rich melts. One successful implementation is the traveling heater method.[49] The perfect quality of crystals grown by this method is accompanied by a low growth rate.

The bulk $Hg_{1-x}Cd_x Te$ crystals are initially used for any types of IR photodetectors. At present these crystals are still used for some IR applications, including n-type single-element photoconductors, SPRITE detectors, and arrays.

3.2.3.3 Epitaxy

The present generation of $Hg_{1-x}Cd_x Te$ is based on epitaxial layers. Compared to bulk growth techniques, epitaxial techniques offer the possibility to grow large-area (≈ 100 cm^2) epilayers and sophisticated layered structures with abrupt and complex composition and doping profiles that can be configured to improve the performance of photodetectors. The growth is performed at low temperatures, which makes it possible to reduce the native defects density. Due to the low Hg pressures, there is no need for thick-walled ampoules and the growth can be carried out in reusable production-type growth systems. The as-grown epilayers can be low-temperature annealed *in situ*. Epilayers can be used for device fabrications without troublesome and time-consuming thinning.

All epitaxial techniques have a common problem: a need for low-cost and large-area substrates that are structurally, chemically, optically, and mechanically matched to the Hg-based semiconductors.[50] No one substrate that satisfies all requirements simultaneously has been found to date.

CdTe and closely lattice-matched ternaries (Cd-Zn-Te, Cd-Te-Se, and Cd-Mn-Te) are the most frequently used substrates that make it possible for epitaxial layers to match the quality standards set by bulk crystals. The substrates are typically grown by the modified vertical and horizontal unseeded Bridgman technique. Most commonly, the (111) and (100) orientations have been used although others, such as (110) and (123), have been tried. The twinning that occurs in the (111) layers can be prevented by a suitable disorientation of the substrate. The limited size, purity problems, Te precipitates, dislocation density, nonuniformity of lattice match, and high price ($60–500 per 1 cm^2 for polished) are remaining problems to be solved. It is believed that these substrates will continue to be important for a long time, particularly for the highest performance devices.

A viable approach to cheap substrates is the use of hybrid substrates, which consist of laminated structures with wafers of bulk crystal, and are covered with buffer lattice-matched layers. Bulk Si, GaAs, and sapphire are some of the high-quality, low cost, and readily available crystals that have been shown to be useful substrates for $Hg_{1-x}Cd_xTe$. The buffer layers are a few micrometers thick of CdTe or (Cd,Zn)Te, obtained *in situ* or *ex situ* with a non-equilibrium growth, typically from the vapor phase. The feasibility of growing high-quality $Hg_{1-x}Cd_xTe$ on hybrid substrates was demonstrated first by Rockwell International. This technology is referred to as Producible Alternative to CdTe for Epitaxy (PACE).[51, 52] The substrates are CdTe/sapphire (PACE 1), CdTe/GaAs (PACE 2) and Si/GaAs/CdZnTe (PACE 3). The use of Si substrates is the most promising for future low-cost devices, which can be readily integrated with Si signal processing electronics.[53–55] Lattice and thermal mismatch, spurious growth, and impurity doping from the substrates are the most serious problems that would require considerable development to meet the stringent requirements of high-performance devices. Large focal planes based on $Hg_{1-x}Cd_xTe$ grown onto (Cd,Zn)Te, sapphire, GaAs, and Si with various epitaxial techniques already have been reported in numerous papers.[1, 4, 56]

Among the epitaxial techniques, the LPE is the most mature method, enabling growth of the device-quality homogeneous layers and multilayered structures.[57] LPE growth can be gained with pseudobinary HgTe-, Te-, and Hg-rich solutions. The first approach has no practical importance at present, mainly because it requires a high temperature ($\approx 700°C$) and high Hg pressure (> 10 atm). Both Te-solution growth (420–500°C) and Hg-solution growth (360–500°C) have been used with equal success in a variety of configurations.

Te-solution growth offers lower Hg pressure (≈ 0.1 atm) and is most frequently used to grow the whole device structures or the base layers only. Perfect undoped and intentionally doped LPE layers were grown on CdTe, (Cd,Zn)Te, Cd(Te,Se), and hybrid substrates.[43, 58–61]

Remaining problems with LPE are the sharpness of the interface region, relatively high density of misfit and threading dislocations, complete Te-rich melt weeping off the grown layers, precipitation of Te upon cooling, low distribution of coefficients for important acceptors As and Sb, frequent occurrences of a terraced surface morphology, and the process of scaling-up to meet industrial needs.

Applications of Hg-rich solution do not present the problem of solution weeping during growth. Another important advantage of using Hg-rich solutions is that they provide well-behaved impurities that can be incorporated with ease. Unlike the growth from Te-rich solutions, the composition of the growing layers from Hg-rich solutions can be drastically changed. Because CdTe in Hg-rich solutions has a low solubility, Cd's depletion of melt is present in Hg-rich solutions. This has been overcome by the use of "infinite-melt" growth developed at the Santa Barbara Research Center (SBRC).[62] The term "infinite melt" refers to the use of a very large (≥ 5 kg) melt that is maintained at a constant temperature and composition for the life of the melt. This technique has been used to grow epilayers on CdZnTe and alternative substrates for large FPAs.[63]

LPE growth from Hg-rich sources is frequently used to grow a ≈ 1 μm P$^+$ wide gap cap epilayer on top of the more thick epilayer prepared by Te-rich source LPE or other methods.[38, 64] Due to a low growth temperature (360–375°C), P$^+$-n junctions are abrupt (<0.1 μm) while heterojunctions are centered at the electrical junction and graded over 0.6–1.3 μm. Because of the depletion of CdTe in the solution, the reduction of the x value toward the layer's surface does not produce problems. Instead, the surface cap layer plays the role of the contact region.

Various vapor phase epitaxy (VPE) methods have been used to grow $Hg_{1-x}Cd_xTe$ layers. One of the oldest is the isothermal vapor phase epitaxy (ISOVPE), which was initially invented in France.[65] ISOVPE is a relatively simple, quasiequilibrium growth technique in which HgTe is transported at a relatively high temperature (400–600°C) from the source (HgTe or $Hg_{1-x}Cd_xTe$) to the substrate by evaporation-condensation mechanisms. An inherent property of the method is in-depth grading because interdiffusion of deposited and substrate materials is involved in layer formation. Homogeneous epilayers can be grown on composite substrates.[66] A reusable growth system has been used for production purposes.[67] The ISOVPE layers can be doped during growth with Au, As, or In.[68]

Negative ISOVPE epitaxy can be used to increase the bandgap at the surface for heterojunction passivation or contacts.[69, 70] With suitable modifications, P-p-n$^+$ and N-n-P$^+$ diode structures can be grown.[68]

Current efforts with VPE are mostly on two nonequilibrium growth methods: metalorganic chemical vapor deposition (MOCVD) and molecular beam epitaxy (MBE). These methods allow for growth at lower temperatures and the capacity for dynamic modification of the growth conditions, which makes it possible to obtain heterostructures with almost any bandgap and doping profiles. The reduced deposition temperature results in a substaintially larger number of possible substrates than those of LPE and ISOVPE. MOCVD is also the growth method that is most frequently used to fabricate the hybrid substrates for $Hg_{1-x}Cd_xTe$.

MOCVD appears to be very promising for large-scale and low-cost production of epilayers.[71–81] Two alternative processes are used: direct alloy growth (DAG) and interdiffused multilayer process (IMP). In IMP growth, the successive layers of CdTe and HgTe have the combined thickness of two consecutive layers that are approximately 0.1 μm. These layers are deposited and interdiffused during growth or during a short annealing *in situ* at the end of the growth. The activation of As, Sb, and In dopants is easier to achieve with IMP when introduced during the CdTe growth cycle and with a Cd/Te flux ratio above 1.[71, 80] Near 100% activation of As was reported for 365°C IMP growth with stoichiometry annealing only.[80] Approximately 50% As activation is achieved at high doping levels for layers subjected to stoichiometric annealing at 250°C, and 100% activation with annealing at 415°C followed by the 250°C anneal.[35, 36, 74, 76, 82] Indium could be used as a standard donor dopant to achieve background doping of the whole MOCVD-grown heterostructure. Iodine is a more stable dopant that can be used to control local doping within a 3×10^{14} to 2×10^{18} cm^{-3} range with 100% activation when following a standard stoichiometric anneal.[35, 76]

Recent reports show that the proper selection of the IMP stages yields the capability to possibly achieve dynamic metal-rich conditions during growth.[36] This enables annihilation of metal vacancies and activation of I and As dopants without any post-growth anneals. Significant orientation dependence of As and I doping was observed for (111B), (211B), and (100).[35, 36, 74, 76, 81, 82]

Intensive studies are currently under way on MBE.[53, 54, 61, 83-90] This technique offers unique capabilities in material and device engineering, including the lowest growth temperature, superlattice growth, and potential for the most complicated composition and doping profiles. While the computerized control of MBE systems is easy, the main drawback of MBE is a high cost of equipment and maintenance. But the quality of the MBE layers is steadily improving. High-quality epilayers have been grown on CdZnTe, GaAs, and Si. Indium is preferable as an n-type dopant. Significant efforts are being spent on As and Sb-doping to improve incorporation during the MBE process and to reduce the temperature required for activation. The metal-saturation conditions cannot be reached at the temperatures required for high-quality MBE growth. The necessity to activate acceptor dopants at high temperatures eliminates the benefits of low-temperature growth. Recently, near 100% activation was achieved for a 2×10^{18} cm^{-3} As concentration with a 300°C activation anneal followed by a 250°C stoichiometric anneal.[88, 91] Efficient p-type doping is being achieved through the combined use of Cd_3As_2 and Te from cracker effusion cells.[92] In situ p-type layers are obtained by flux manipulation. However, these layers have only partial As activation.[93] Reduced growth temperatures (175°C), Cd_3As_2 effusion cells, the interdiffused superlattice process of doping during the CdTe deposition cycle, photo-assisted MBE, As planar doping, and the use of Te_2 and As_4 crackers are some of the ways to improve As incorporation.

The MBE-grown layers have been used to fabricate high-quality IR devices. But several problems still exist, including twin formations, the requirement of very good surface preparation prior to growth, uncontrolled doping (10^{15} cm^{-3}), dislocation density, and composition inhomogenities.

While recent remarkable achievements have improved IR devices, the further development of MBE is necessary to fully apply this technology to IR sensors. One attractive new growth technique combines MOCVD and MBE: metalorganic molecular beam epitaxy.[94, 95] With this new development, significant activation has been achieved without any annealing.

3.3 Hg-based Alternatives to HgCdTe

At present HgCdTe is the most important intrinsic semiconductor alloy system for IR detectors. However, HgCdTe is one of the most difficult materials to use for IR detectors. Currently, HgCdTe FPAs are limited by the yield of arrays, which increases their cost. In spite of achievements in material and device quality, difficulties still exist due to lattice, surface, and interface instabilities. This realization—when combined with continued progress in the growth of new ternary alloy systems and artificial semiconductor heterostructures—has intensified the search for alternative IR materials.

Among the small-gap II-VI semiconductors for IR detectors, only Cd, Zn, Mn, and Mg have been shown to open the bandgap of the Hg-based binary semimetals HgTe and HgSe to match the IR wavelength range. It appears that the amount of Mg to introduce in HgTe to match the 10-μm range is insufficient to reinforce the Hg-Te bond.[96] The main obstacles to the technological development of $Hg_{1-x}Cd_xSe$ are the difficulties in obtaining type conversion. From the above alloy systems, $Hg_{1-x}Zn_xTe$ (HgZnTe) and $Hg_{1-x}Mn_xTe$ (HgMnTe) occupy a privileged position.

Neither HgZnTe nor HgMnTe have ever been systematically explored in the device context. There are several reasons for this. First, preliminary investigations of these alloy systems came on the scene after HgCdTe detector development was well on its way. Second, the HgZnTe alloy presents a more serious technological challenge than HgCdTe. In the case of HgMnTe, Mn is not a group II element, and as a result HgMnTe is not truly a II-VI alloy. This ternary compound was viewed with some suspicion by those not directly familiar with its crystallographic, electrical, and optical behavior. In such a situation, proponents of the parallel development of HgZnTe and HgMnTe for IR detector fabrication encountered considerable difficulty in selling the idea to industry and funding agencies.

In 1985, Sher et al. showed from theoretical consideration that the weak HgTe bond is destabilized by alloying it with CdTe, but stabilized by ZnTe.[97] Many groups worldwide have become very interested in this prediction, and more specifically, in the growth and properties of the HgZnTe alloy system as the material for photodetection application in the IR spectral region. But the question of lattice stability in the case of HgMnTe compound is rather ambiguous. According to Wall et al., the Hg-Te bond stability of this alloy is similar to that observed in the binary narrow-gap parent compound.[98] This conclusion contradicts more recently published results.[99] It has been established that the incorporation of Mn in CdTe destabilizes its lattice because of the Mn 3d orbitals hybridizing into the tetrahedral bonds.[100] This section reviewed only selected topics on the growth process and physical properties of HgZnTe and HgMnTe ternary alloys. More information can be found in two comprehensive reviews—Refs. 101 and 102—and the books cited in Refs. 1 and 103.

3.3.1 Crystal growth

The pseudobinary diagram for HgZnTe is responsible for serious problems encountered in crystal growth, including:

- The separation between the liquidus and solidus curves are large and lead to high segregation coefficients;
- The solidus lines that are flat result in a weak variation of the growth temperature that causes a large composition variation; and
- The very high Hg pressure over melts makes the growth of homogeneous bulk crystals quite unfavorable.

For comparison, Fig 3.3(a) shows HgTe-ZnTe and HgTe-CdTe pseudobinary phase diagrams.

HgTe and MnTe are not completely miscible over the entire range, but the $Hg_{1-x}Mn_xTe$ single-phase region is limited to approximately $x < 0.35$.[104] As discussed by Becla et al., the solidus-liquidus separation in the pseudo-binary HgTe-MnTe system is more than two times narrower than in the corresponding HgTe-CdTe system.[105] This conclusion has been confirmed by Bodnaruk et al.[106] Consequently, to meet the same demand for cut-off wavelength homogeneity, the HgMnTe crystals must be much more uniform than similarly grown HgCdTe crystals.

For the growth of bulk HgZnTe and HgMnTe single crystals, three methods are currently the most popular: Bridgman-Stockbarger, solid state recrystallization, and the travelling heater method (THM). The best quality HgZnTe crystals have been produced by THM. Using this method, Triboulet et al.[107] produced $Hg_{1-x}Zn_xTe$ crystals ($x \approx 0.15$) with a longitudinal homogeneity of ± 0.01 mol and radial homogeneity of ± 0.01 mol. The source material was a cylinder composed of two cylindrical segments—one HgTe, the other ZnTe—the cross section of which was in the ratio corresponding to the desired composition.

To improve the crystalline quality of HgMnTe single crystals, different modified techniques have been used. Gille et al.[108] demonstrated $Hg_{1-x}Mn_xTe$ ($x \approx 0.10$) single crystals grown by THM with standard deviation $\Delta x = \pm 0.003$ along a 16-mm-diameter slice of crystals. Becla et al.[105] decreased the radial macrosegregation and eliminated small-scale compositional undulations in the vertical Bridgman-grown material by applying a 30-kG magnetic field. Takeyama and Narira[109] developed an advanced crystal growth method called the modified two-phase mixture method to produce highly homogeneous, large single crystals of ternary and quaternary alloys.

The best performance of modern devices, however, requires more sophisticated structures. These structures are only achieved by using epitaxial growth techniques. Additionally, when compared to bulk growth techniques, epitaxial techniques offer important advantages, including lower temperatures and Hg vapor pressures, shorter growth times, and reduced precipitation problems that enable the growth of large-area samples with good lateral homogeneity. The above advantages have prompted research in a variety of thin-film growth techniques, such as VPE, LPE, MBE, and MOCVD. The first studies of LPE crystal growth of HgCdZnTe and HgCdMnTe from Hg-rich solutions have demonstrated that the homogeneity of epilayers can be improved by incorporating Zn or Mn during the crystal growth.[110] More recently, considerable progress has been achieved in HgMnTe film fabrication by MOCVD using an inderdiffused multilayer process.[111] Depending on growth conditions, both n- and p-type layers may be produced with extrinsic electron and hole concentrations of the order of 10^{15} cm^{-3} and 10^{14} cm^{-3}, respectively.

All of the epitaxial growth processes depend on the identification of suitable substrates. They require large-area, single-crystal substrates. The large difference in lattice parameters of HgTe and ZnTe induces strong interactions between cations. Vegard's law appears to be obeyed relatively better in HgZnTe than in HgCdTe, and for $Hg_{1-x}Zn_xTe$ at 300 K $a(x) = 6.461 - 0.361x$ (Å).[103] In comparison with HgCdTe, the lattice parameter of $Hg_{1-x}Mn_xTe$ $a(x) = 6.461 - 0.121x$ (Å) varies with x much more rapidly, which is a disadvantage from the point of view of the epitaxial growth of multilayer heterostructures that is required for advanced IR devices. The lattice parameter of the zinc-blende compound $Cd_{1-x}Zn_xTe$ ($Cd_{1-x}Mn_xTe$) indicates a simple matter: to find suitable substrates for epitaxial growth of $Hg_{1-x}Zn_xTe$ ($Hg_{1-x}Mn_xTe$). However, Bridgman-grown CdMnTe crystals are highly twinned and thus unusable as epitaxial substrates.

3.3.2 Physical properties

The physical properties of both ternary alloys are determined by the energy gap structure near the Γ-point of the Brillouin zone. The shape of the electron band and the light mass hole band is determined by the k·p theory. The bandgap structure of HgZnTe near the Γ-point is similar to that of the HgCdTe ternary illustrated in Fig. 3.1. The bandgap energy of HgZnTe varies approximately 1.4 times (2 times for HgMnTe) as fast with the composition parameter x as it does for HgCdTe.

Both HgZnTe and HgMnTe exhibit compositional-dependent optical and transport properties similar to HgCdTe materials with the same energy gap. Some physical properties of alternative alloys indicate a structural advantage in comparison with HgCdTe. Introducing ZnTe in HgTe decreases statistically the ionicity of the bond, improving the stability of the alloy. Moreover, because the bond length of ZnTe (2.406 Å) is 14% shorter than that of HgTe (2.797 Å) or CdTe (2.804 Å), the dislocation energy per unit length and the hardness of the HgZnTe alloy are higher than that of HgCdTe. The maximum degree of microhardness for HgZnTe is more than twice that for HgCdTe.[112] HgZnTe is a material that is more resistant to dislocation formation and plastic deformation than HgCdTe.

The as-grown $Hg_{1-x}Zn_xTe$ material is highly p-type in the 10^{17} cm^{-3} range with mobilities in the hundreds of cm^2/V s. These values indicate that its conduction is dominated by holes arising from Hg vacancies. After a low-temperature anneal ($T \leq 300°C$) accomplished with an excess of Hg (which annihilates the

Hg vacancies), the material is converted to low n-type in the mid 10^{14} cm^{-3} to low 10^{15} cm^{-3} range, which has mobilities ranging from 10^4 to 4×10^5 cm^2/V s. A study by Rolland et al.[113] showed that the n-type conversion occurs only for crystals with composition $x \leq 0.15$. The Hg diffusion rate is slower in HgZnTe than in HgCdTe. Interdiffusion studies between HgTe and ZnTe indicate that the interdiffusion coefficient is approximately 10 times lower in HgZnTe than in HgCdTe.[107]

Berding et al.[114] and Granger et al.[115] gave the theoretical description of the scattering mechanisms in HgZnTe. To obtain a good fit to experimental data for Hg$_{0.866}$Zn$_{0.134}$Te, they considered phonon dispersion plus ionized impurity scattering plus core dispersion without compensation in their mobility calculations. Theoretical calculations of electron mobilities[107] indicate that the disorder scattering is negligible for HgZnTe alloy. In contrast, the hole mobilities are likely to be limited by alloy scattering, and the predicted alloy hole mobility of HgZnTe is approximately a factor of 2 less than what was found for HgCdTe. Additionally, Abdelhakiem et al.[116] confirmed that the electron mobilities are very close to HgCdTe ones for the same energy gap and the same donor and acceptor concentrations.

The HgMnTe alloy is a semimagnetic narrow-gap semiconductor. The exchange interaction between band electrons and Mn^{2+} electrons modifies their band structure, making it dependent on the magnetic field at very low temperature. In the range of temperatures typical for IR detector operation ($\geq$77 K), the spin-independent properties of HgMnTe are practically identical to the properties of HgCdTe, which are discussed exhaustively in the literature. The studies carried out by Kremer et al.[117] confirmed that the annealing of samples in Hg vapor eliminates the Hg vacancies, with the resulting material being n-type due to some unknown native donor. The diffusion rate of Hg into HgMnTe is the same as into HgCdTe. Measurements of the transport properties of Hg$_{1-x}$Mn$_x$Te ($0.095 \leq x \leq 0.15$) indicate deep donor and acceptor levels into the energy gap, which influence not only the temperature dependencies of the Hall coefficient, conductivity, and Hall mobility, but also the minority carrier lifetimes.[118, 119] Theoretical considerations of the electron mobilities in HgCdTe and HgMnTe indicate that at room temperature the mobilities are nearly the same. But at 77 K, the electron mobilities are approximately 30% less for HgMnTe when compared with the same concentration of defects.[120]

The measured carrier lifetime in both ternary alloy systems is a sensitive characteristic of semiconductors that depends on material composition, temperature, doping, and defects. The Auger mechanism governs the high-temperature lifetime, and the Shockley-Read mechanism is mainly responsible for low-temperature lifetimes. The reported positions of SR centers for both n- and p-type materials range anywhere from near the valence to near the conduction band. Comprehensive reviews of generation-recombination mechanisms and the carrier lifetime experimental data for both ternary alloys are given by Rogalski et al.[1, 103]

Tables 3.4 and 3.5 contain lists of standard approximate relationships for material properties of HgZnTe and HgMnTe, respectively. Most of these relationships have been taken from Refs. 101 and 102. Some of these parameters, e.g., the intrinsic carrier concentration, have since been re-examined. For example, Sha et al.[121]

Table 3.4 Standard relationships for $Hg_{1-x}Zn_xTe$ ($0.10 \leq x \leq 0.40$). Data were obtained from Ref. 101.

Parameter	Relationship
Lattice constant $a(x)$ (nm) at 300 K	$0.6461 - 0.0361x$
Density γ (g/cm^3) at 300 K	$8.05 - 2.41x$
Energy gap E_g (eV)	$-0.3 + 0.0324x^{1/2} + 2.731x - 0.629x^2 + 0.533x^3$ $+ 5.3 \times 10^{-4}T(1 - 0.76x^{1/2} - 1.29x)$
Intrinsic carrier concentration n_i (cm^{-3})	$(3.607 + 11.370x + 6.584 \times 10^{-3}T - 3.633 \times 10^{-2}xT)$ $\times 10^{14}E_g^{3/4}T^{3/2}\exp(-5802E_g/T)$
Momentum matrix element P (eV cm)	8.5×10^{-8}
Spin-orbit splitting energy Δ (eV)	1.0
Effective masses: m_e^*/m $\quad\quad\quad\quad m_h^*/m$	$5.7 \times 10^{-16}E_g/P^2$ E_g in eV; P in eV cm 0.6
Mobilities: μ_e (cm^2/V s) $\quad\quad\quad \mu_h$ (cm^2/V s)	$9 \times 10^8 b/T^{2a}$ $a = (0.14/x)^{0.6}$; $b = (0.14/x)^{7.5}$ $\mu_e(x,T)/100$
Static dielectric constant ε_s	$20.206 - 15.153x + 6.5909x^2 - 0.951826x^3$
High-frequency dielectric constant ε_∞	$13.2 + 19.1916x + 19.496x^2 - 6.458x^3$

Table 3.5 Standard relationships for $Hg_{1-x}Mn_xTe$ ($0.08 \leq x \leq 030$). Data were obtained from Ref. 102.

Parameter	Relationship
Lattice constant $a(x)$ (nm) at 300 K	$0.6461 - 0.0121x$
Density γ (g/cm^3) at 300 K	$8.12 - 3.37x$
Energy gap E_g (eV)	$-0.253 + 3.446x + 4.9 \times 10^{-4}xT - 2.55 \times 10^{-3}T$
Intrinsic carrier concentration n_i (cm^{-3})	$(4.615 - 1.59x + 2.64 \times 10^{-3}T - 1.70 \times 10^{-2}xT$ $+ 34.15x^2) \times 10^{14}E_g^{3/4}T^{3/2}\exp(-5802E_g/T)$
Momentum matrix element P (eVcm)	$(8.35 - 7.94x) \times 10^{-8}$
Spin-orbit splitting energy Δ (eV)	1.08
Effective masses: m_e^*/m $\quad\quad\quad\quad m_h^*/m$	$(8.35 - 7.94x) \times 10^{-8}E_g$ in eV; P in eV cm 0.5
Mobilities: μ_e (cm^2/V s) $\quad\quad\quad \mu_h$ (cm^2/V s)	$9 \times 10^8 b/T^{2a}$ $a = (0.095/x)^{0.6}$; $b = (0.095/x)^{7.5}$ $\mu_e(x,T)/100$
Static dielectric constant ε_s	$20.5 - 32.6x + 25.1x^2$
High-frequency dielectric constant ε_∞	$15.2 - 28x + 28.2x^2$

concluded that their improved calculations of intrinsic carrier concentration were approximately 10–30% higher than those obtained earlier by Jóźwikowski and Rogalski.[122] However, the new calculations also should be treated as approximations since the dependence of the energy gap on composition and temperature $[E_g(x, T)$ is necessary in the calculations of $n_i]$ is still under serious discussion.[123]

3.4 InAs/Ga$_{1-x}$In$_x$Sb Type II Superlattices

An InAs/Ga$_{1-x}$In$_x$Sb (InAs/GaInSb) type II strained layer superlattice (SLS) can be considered as an alternative to a HgCdTe material system. It is characterized by strong absorption (high quantum efficiency) and low thermal generation rate. Consequently, SLS structures provide high responsivity, which already has been reached with HgCdTe material. A further advantage of type-II SLS has been well established in III-V process technology.

The InAs/GaInSb material system is, however, in a very early stage of development. Problems exist in material growth, processing, substrate preparation, and device passivation.[124] Optimization of SL growth is a trade-off between the interfaces' roughness (which results in smoother interfaces at higher temperatures) and residual background carrier concentrations (which are minimized on the low end of this range). The thin nature of InAs and GaInSb layers (<8 nm) necessitates low growth rates to control each layer's thickness to within one (or one-half) monolayer (ML). Monolayer fluctuations of the InAs layer thickness can shift the cut-off wavelength by approximately ± 2 μm for a 20-μm designed cut-off. Typical growth rates are less than 1 MLs for each layer.

The type II SL has a staggered band alignment, resulting in a conduction band of the InAs layer that is lower than the valence band of the InGaSb layer, as shown in Fig. 3.5. This creates a situation in which the energy bandgap of the SL can be adjusted to form either a semimetal (for wide InAs and GaInSb layers) or a narrow bandgap (for narrow layers) semiconductor material. The type II band alignment and internal strain lowers the conduction band minimum of InAs and raises the heavy-hole band in Ga$_{1-x}$In$_x$Sb by the deformation potential effect. This reduced bandgap is advantageous because longer cut-off wavelengths can be obtained with

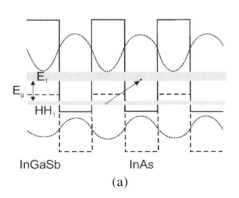
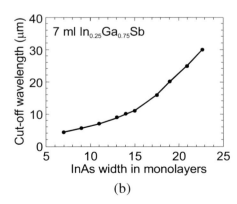

(a)   (b)

Figure 3.5 InAs/GaInSb strained layer superlattice: (a) band edge diagram illustrating the confined electron and hole minibands that form the energy bandgap; (b) change in cut-off wavelength with change in one SL parameter, the InAs layer width. (Reprinted from Ref. 125.)

a reduced layer thickness in the strained SL, leading to an optical absorption coefficient comparable to that of HgCdTe. The hole wave function has less overlap with neighboring confined hole states than the electron wave function has with neighboring electron states. In the SL, the electrons are mainly located in the InAs layers, whereas holes are confined to the GaInSb layers. Optical transitions occur spatially indirectly; as a result, the optical matrix element for such transitions is relatively small. The bandgap of the SL is determined by the energy difference between the electron miniband E_1 and the first heavy-hole state HH_1 at the Brillouin zone center. It can be varied continuously in a range from 0 to approximately 250 meV. An example of the wide tunability of the SL is shown in Fig. 3.5(b). The dispersion of the heavy-hole states exhibits a strong anisotropy while the other subbands are nearly isotropic.[124]

In addition to the capability of adjusting the bandgap energy, InAs/GaInSb SL material provides the added capability of energy band structure engineering. It appears that the same bandgap energy can be designed using different combinations of layer thickness and composition. By examining the band structure of these equivalent designs, the optimum design can be selected to enhance important parameters such as IR absorption coefficient, electron effective mass, and charge carrier lifetime. For example, the SL energy bands can be structured so that there is a larger energy separation between the heavy- and light-hole bands than the SL bandgap energy. This suppresses Auger recombination mechanisms and thereby enhances carrier lifetime.

It has been suggested that InAs/Ga$_{1-x}$In$_x$Sb SLSs material systems can have some advantages over bulk HgCdTe, including lower leakage currents and greater uniformity.[126, 127] According to Smith and Mailhiot (Ref. 126), the electronic properties of SLSs may be superior to those of the HgCdTe alloy. The effective masses are not directly dependent on the bandgap energy, as is the case for a bulk semiconductor.[128] The electron effective mass of InAs/GaInSb SLS is larger ($m^*/m_o \approx 0.02$–0.03, compared to $m^*/m_o = 0.009$ in HgCdTe alloy with the same bandgap $E_g \approx 0.1$ eV).[129] Thus, diode tunneling currents in the SL can be reduced in comparison with the HgCdTe alloy. Although in-plane mobilities drop precipitously for thin wells, electron mobilities approaching 10^4 cm^2/V s have been observed in InAs/GaInSb SLs with the layers less than 40 Å thick. While mobilities in these SLs are found to be limited by the same interface roughness scattering mechanism, detailed band structure calculations reveal a much weaker dependence on layer thickness, in reasonable agreement with experimental results.[130]

A consequence of the type II band alignment of an InAs/GaInSb material system is the spatial separation of electrons and holes. This is particularly disadvantageous for optical absorption where a significant overlap of electron and hole wave function is needed. However, a reduction in the electronic confinement can be achieved by growing thinner GaInSb barriers or by introducing more In into the GaInSb layers, which leads to an optical absorption coefficient comparable to that of HgCdTe.

Theoretical analysis of band-to-band Auger and radiative recombination lifetimes for InAs/GaInSb SLSs has shown that in these objects, the p-type Auger recombination rates are suppressed by several orders when compared to those of bulk
HgCdTe with similar bandgaps, but n-type materials are less advantageous.[131, 132]
However, the promise of Auger suppression has yet to be observed in practical
device material.

A comparison of theoretically calculated and experimentally observed lifetimes at 77 K for 10 μm InAs/GaInSb SLS and 10 μm HgCdTe is presented
in Fig. 3.6. The agreement between theory and experiment for carrier densities
above 2×10^{17} cm^{-3} is good. The discrepancy between both types of results for
lower carrier densities is due to Shockley-Read recombination processes having a
$\tau \approx 6 \times 10^{-9}$ s, which was not taken into account in the calculations. For higher
carrier densities, the SL carrier lifetime is two orders of magnitude longer than in
HgCdTe. However, in low doping regions (below 10^{15} cm^{-3}, which are necessary
to fabricate high-performance p-on-n HgCdTe photodiodes), the experimentally
measured carrier lifetime in HgCdTe is more than two orders of magnitude longer
than in SL. The recently published upper experimental data in Ref. 134 coincides
well with HgCdTe trends in the range of lower carrier concentrations (see Fig. 3.6).
In general, however, the SL carrier lifetime is limited by the influence of trap centers located at an energy level of ~1/3 bandgap below the effective conduction
band edge.

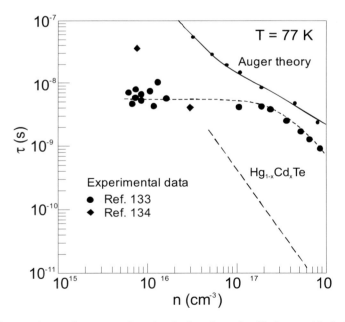

Figure 3.6 Comparison of measured and calculated carrier lifetimes of InAs/GaInSb SLS
(approximately 120 eV energy gap) at 77 K as a function of carrier density. The experimental
data were taken from Refs. 133 (●) and 134 (◆), and the theoretical data were taken from
Ref. 133.

Narrow bandgap materials require the doping to be controlled to at least 1×10^{15} cm^{-3} or below to avoid deleterious high-field tunneling currents across reduced depletion widths at temperatures below 77 K. Lifetimes must be increased to enhance carrier diffusion and reduce related dark currents. At the present stage of development, the residual doping concentration (both n-type as well as p-type) is above 6×10^{15} cm^{-3} in SLs grown at a substrate temperature ranging from 360°C to 440°C.[124]

3.5 Novel Sb-Based Materials

In the middle and late 1950s, it was discovered that InSb, a member of the newly discovered III-V compound semiconductor family, had the smallest energy gap of any semiconductor known at that time; its potential applications as a MWIR detector became obvious. The energy gap of InSb is less well matched to the 3–5-μm band at higher operating temperatures, and better performance can be obtained from Hg$_{1-x}$Cd$_x$Te. InAs is a similar compound to InSb but has a larger energy gap, so its threshold wavelength is 3–4-μm.

3.5.1 InAsSb

One of the earliest InSb-based ternary alloys investigated was InAs$_{1-x}$Sb$_x$. The properties of the InAsSb ternary alloy were first investigated by Woolley et al.[135, 136] This material has a zinc-blende structure and direct bandgap at the Brillouin zone center. The relationship between the energy gap and composition is illustrated in Fig. 3.7, and it may be described by the following expression:

$$E_g(x, T) = 0.411 - \frac{3.4 \times 10^{-4} T^2}{210 + T} - 0.876x + 0.70x^2$$
$$+ 3.4 \times 10^{-4} x T (1 - x).^{[137]} \qquad (3.10)$$

This formula indicates a fairly weak dependence of the band edge on composition in comparison with HgCdTe. As can be seen in Fig. 3.7, there is a strong bowing and a room temperature minimum bandgap energy of 0.1 eV at 35%.

Earlier data suggest that InAsSb is an attractive semiconductor material for detectors covering the 3–5-μm and 8–12-μm spectral ranges.[103] However, some recent experimental results demonstrated that the cutoff wavelength of epitaxial layers can be longer than 12.5 μm, thus covering the entire 8–14 μm at near room temperature. This may be due to structural ordering, but the exact mechanism has not been determined yet. These results show that InAsSb is a promising material system for 8–14-μm photodetectors operating at near room temperatures.[138, 139]

Progress in the development of InAsSb material fabrication has been limited by crystal synthesis problems. The large separation between the liquidus and solidus

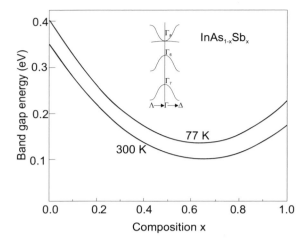

Figure 3.7 Dependence of the bandgap energy of $InAs_{1-x}Sb_x$ on composition.

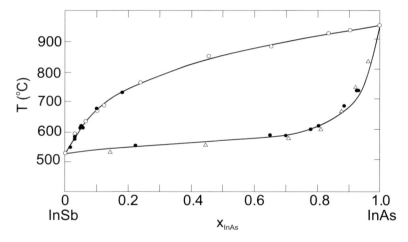

Figure 3.8 Pseudobinary phase diagram for the InAs-InSb system. (Reprinted from Ref. 140 with permission from ECS.)

(Fig. 3.8) and the lattice mismatch (6.9% between InAs and InSb) place stringent demands upon the method of crystal growth. These difficulties are being overcome systematically using MBE and MOCVD.[141, 142]

Although effort has been devoted to the development of InAsSb as an alternative to HgCdTe for IR applications, there is relatively little information available on the physical properties of InAsSb. A III–V detector technology would benefit from superior bond strengths and material stability (compared to HgCdTe), well-behaved dopants, and high-quality III–V substrates. Woolley and co-workers established the InAs-InSb miscibility,[143] the pseudobinary phase diagram,[135] scattering mechanisms,[144] and the dependence of fundamental properties such as bandgap[136] and effective masses on composition.[145] All of the above measurements were performed on polycrystalline samples prepared by various freezing and annealing techniques. The review of the rapid development of InAsSb crystal growth techniques, physical properties, and detector fabrication procedures is presented in Refs. 103 and 146.

Conventional InAsSb does not have a sufficiently small gap at 77 K for operation in the 8–14-μm wavelength range. To build detectors that span the 8–14 μm atmospheric window, Osbourn proposed the use of InAsSb SLSs.[147] This theoretical prediction reinforced interest in the investigation of InAsSb ternary alloy as a material for IR detector applications.[1, 146] However, further development of InAsSb SLSs has failed due to difficulties encountered in finding the proper growth conditions, especially for SLSs in the middle region of composition. Control of alloy composition has been problematic, especially for MBE. Due to the spontaneous nature of CuPt orderings, which result in substantial bandgap shrinkage, it is difficult to accurately and reproducibly control the desired bandgap for optoelectronic device applications.[148]

3.5.2 InTlSb and InTlP

Since an $InAs_{0.35}Sb_{0.65}$ based detector is not sufficient for efficient IR detection operated at lower temperatures in the 8–12 μm range, $In_{1-x}Tl_xSb$ (InTlSb) was proposed as a potential IR material in the LWIR region.[149, 150] TlSb is predicted as a semimetal. By alloying TlSb with InSb, the bandgap of InTlSb could be varied from −1.5 eV to 0.26 eV. Assuming a linear dependence of the bandgap on alloy composition, $In_{1-x}Tl_xSb$ can then be expected to reach a bandgap of 0.1 eV at $x = 0.08$, while exhibiting a similar lattice constant as InSb since the radius of Tl atom is very similar to In. At this gap, InTlSb and HgCdTe have very similar band structure. This implies that InTlSb has comparable optical and electrical properties to HgCdTe. In the structural aspect, InTlSb is expected to be more robust due to stronger bonding. The estimated miscibility limit of Tl in zinc-blende InTlSb was estimated to be approximately 15%, which is sufficient to obtain an energy gap down to 0.1 eV. Figure 3.9 shows the expected relationships between bandgap energy and lattice constant for Tl-based III-V zinc-blende alloys.[151]

Room-temperature operation of InTlSb photodetectors has been demonstrated with a cutoff wavelength of approximately 11 μm.[142]

Van Schilfgaarde et al.[152] showed that another ternary alloy, $In_{1-x}Tl_xP$ (InTlP), is a promising material for IR detectors. It was shown that this material can cover the bandgap from 1.42 eV (InP at 0 K) to 0 eV using a small lattice mismatch with InP. Optical measurements verified the reduction of the bandgap by the addition of Tl into the InP.[153]

Tl-based III-V alloys will be more widely studied and applied to devices if the difficulties in the crystal growth are overcome.

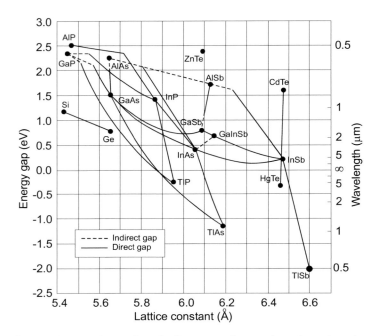

Figure 3.9 Composition and wavelength diagrams of a number of semiconductors with diamond and zinc-blende structure versus their lattice constants. Tl-based III-Vs are also included.

3.5.3 InSbBi

As another alternative to the HgCdTe material system, $InSb_{1-x}Bi_x$ (InSbBi) has been considered because the incorporation of Bi into InSb produces a rapid reduction in the bandgap of 36 meV/%Bi. Thus, only a few percent of Bi is required to reduce the bandgap energy.

The growth of an InSbBi epitaxial layer is difficult due to the large solid-phase miscibility gap between InSb and InBi. Recently, the successful growth of an InSbBi epitaxial layer on InSb and GaAs (100) substrates with a substantial amount of Bi ($\sim$5%) was demonstrated using low-pressure MOCVD.[154, 155] The responsivity of the $InSb_{0.95}Bi_{0.05}$ photoconductor at 10.6 µm was 1.9×10^{-3} V/W at room temperature, and the corresponding Johnson noise limited detectivity was 1.2×10^6 cm Hz$^{1/2}$/W. The effective carrier lifetime estimated from bias voltage-dependent responsivity was approximately 0.7 ns at 300 K.

3.5.4 InSbN

Dilute nitride alloys of III-V semiconductors have progressed rapidly in recent years following the discovery of strong negative bandgap bowing effects.[156, 157] Most papers concentrated on the alloys GaAsN and GaInAsN owing to their technological importance for fiber communications at wavelengths of 1.3 and 1.55 µm.

Initial estimates indicated that the addition of N to InSb would lead to a reduction of its bandgap at a rate similar to that of the wider-gap III-V materials of approximately 100 meV per percent of N. Therefore, for some applications the InSbN might provide an alternative that would overcome some of the limitations of the more established materials such as HgCdTe and type II SLs of InAs/GaInSb.

Preliminary estimates of the band structure of $InSb_{1-x}N_x$ were made using a semi-empirical $k \cdot p$ model. The predicted variation in band structure is illustrated in Fig. 3.10. A decrease in bandgap of 110 meV (fractional change of 63%) was predicted at 1% N, which clearly offers potential for long-wavelength applications. These theoretical predictions were experimentally confirmed with measurements of response wavelengths of light-emitting diodes. The $InSb_{1-x}N_x$ samples with up to 10% of the N were grown by combining molecular beam epitaxy and a nitrogen plasma source.

The bandgap reduction has been accompanied by an enhancement in the Auger recombination lifetime of a factor of approximately 3 in comparison with an equivalent HgCdTe bandgap, due to the higher electron mass and conduction band nonparabolicity.[159]

3.6 Lead Salts

The properties of the lead salt binary and ternary alloys have been extensively reviewed.[1, 4, 7, 42, 160–164] Therefore, only some of their most important properties will be mentioned here.

Lead salt detectors were developed during World War II by the German military for use as heat-seeking sensors to find weapons. Immediately after the war, communications, fire control, and search system applications began to stimulate a strong

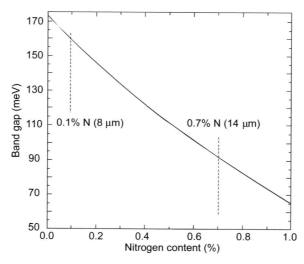

Figure 3.10 Variation of the bandgap energy of $InSb_{1-x}N_x$ over the mid- to long-wavelength IR spectral range. (Reprinted from Ref. 158 with permission from Elsevier.)

development effort that continues to the present day. Sidewinder heat-seeking IR-guided missiles received a great deal of public attention. After 60 years, low-cost, versatile PbS and PbSe polycrystalline thin films remain the photoconductive detectors of choice for many applications in the 1–3 and 3–5 μm spectral regions. Over the past 25 years, new markets have developed for commercial applications of these detectors. Some of the commercial applications include spectrometry, protein analysis, fire detection systems, combustion control, and moisture detection and control.

3.6.1 Physical properties

Lead chalcogenide semiconductors have a face-cantered cubic (rock salt) crystal structure and hence obtained the name "lead salts." Thus, they have [100] cleavage planes and tend to grow in the (100) orientation although they can also be grown in the [111] orientation. Only SnSe possesses the orthorhombic-B29 structure.

Numerous techniques to prepare lead salt single crystals and epitaxial layers have been investigated, and several excellent review articles devoted to this topic have been published.[165–167] Bridgman-type or Czochralski methods result in crystals with increased size and variable composition. Frequently, the material also shows inclusions and rather high dislocation densities. They are mainly used as substrates for the subsequent growth of epilayers. Growth from the solution and the traveling solvent method offer interesting advantages such as higher homogeneity in composition and lower temperatures, which lead to lower concentrations of lattice defects and impurities. The best results have been achieved with the sublimation growth technique since lead salts sublime as molecules.

Thin single-crystal films of IV-VI compounds have found broad application in both fundamental research and applications. The epitaxial layers are usually grown by either VPE or LPE techniques. The best-quality devices have been obtained using MBE.[167]

Lead salts can exist with very large deviations from stoichiometry, and it is difficult to prepare material with carrier concentrations below approximately 10^{17} cm^{-3}. The width of the solidus field is large in IV-VI compounds ($\approx 0.1\%$) making the doping by native defects very efficient. Deviations from stoichiometry create n- or p-type conduction. Native defects associated with excess metal (nonmetal vacancies or possibly metal interstitials) yield acceptor levels, while those that result from excess nonmetal (metal vacancies or possibly nonmetal interstitials) yield donor levels.

In crystals grown from high-purity elements, the effects of foreign impurities are usually negligible when the carrier concentration, due to lattice defects, is above 10^{17} cm^{-3}. Below this concentration, foreign impurities can play a role by compensating for lattice defects and other foreign impurities.

Lead salts have direct energy gaps, which occur at the Brillouin zone edge at the L point. The effective masses are therefore higher and the mobilities lower than for a zinc-blende structure with the same energy gap at the Γ point (the zone center).

The constant-energy surfaces are ellipsoids characterized by the longitudinal and transverse effective masses m_l^* and m_t^*, respectively. The anisotropy factor for PbTe is of the order of 10. The factor is much less, approximately 2, for PbS and PbSe.

Table 3.6 contains a list of material parameters for different types of binary lead salts.

As a consequence of the similar valence and conduction bands of lead salts, the electron and hole mobilities are approximately equal for the same temperatures and doping concentrations. Room-temperature mobilities in lead salts are 500–2000 cm^2/V s. In many high-quality single-crystal samples, the mobility due to lattice scattering varies as $T^{-5/2}$.[161] This behavior has been attributed to a combination of polar-optical and acoustical lattice scattering and achieves the limiting values in the range of 10^5–10^6 cm^2/V s due to defect scattering.

The interband absorption of the lead salts is more complicated compared to the standard case due to the anisotropic multivalley structure of both conduction and valence bands, nonparabolic Kane-type energy dispersion, and k-dependent matrix

Table 3.6 Physical properties of lead salts. Data were obtained from Refs. 160 and 165.

	T (K)	PbTe	PbSe	PbS
Lattice structure		Cubic (NaCl)	Cubic (NaCl)	Cubic (NaCl)
Lattice constant a (nm)	300	0.6460	0.61265	0.59356
Thermal expansion coefficient	300	19.8	19.4	20.3
$\quad \alpha$ (10^{-6} K^{-1})	77	15.9	16.0	
Heat capacity C_p (J mol^{-1} K^{-1})	300	50.7	50.3	47.8
Density γ (g/cm^3)	300	8.242	8.274	7.596
Melting point T_m (K)		1197	1354	1400
Bandgap E_g (eV)	300	0.31	0.28	0.42
	77	0.22	0.17	0.31
	4.2	0.19	0.15	0.29
Thermal coefficient of E_g	80–300	4.2	4.5	4.5
$\quad$ (10^{-4} eV K^{-1})				
Effective masses				
$\quad m_{et}^*/m$	4.2	0.022	0.040	0.080
$\quad m_{ht}^*/m$		0.025	0.034	0.075
$\quad m_{el}^*/m$		0.19	0.070	0.105
$\quad m_{hl}^*/m$		0.24	0.068	0.105
Mobilities				
$\quad \mu_e$ (cm^2/V s)	77	3×10^4	3×10^4	1.5×10^4
$\quad \mu_h$ (cm^2/Vs)		2×10^4	3×10^4	1.5×10^4
Intrinsic carrier concentration n_i	77	1.5×10^{10}	6×10^{11}	3×10^7
$\quad$ (cm^{-3})				
Static dielectric constant ε_s	300	380	206	172
	77	428	227	184
High-frequency dielectric	300	32.8	22.9	17.2
$\quad$ constant ε_∞				
	77	36.9	25.2	18.4
Optical phonons				
$\quad$ LO (cm^{-1})	300	114	133	212
$\quad$ TO (cm^{-1})	77	32	44	67

elements. Analytical expressions for the absorption coefficient for energies near the absorption edge have been given by several researchers.[168–170]

3.6.2 Deposition of polycrystaline PbS and PbSe films

Although the fabrication methods developed for these photoconductors are not completely understood, their properties are well established. Unlike most other semiconductor IR detectors, lead salt materials are used in the form of polycrystalline films approximately 1 µm thick and with individual crystallites ranging in size from approximately 0.1–1.0 µm. They are usually prepared by chemical deposition using empirical recipes, which generally yields better uniformity of response and more stable results than the evaporative methods.[171–175]

The PbSe and PbS films used in commercial IR detectors are made by chemical bath deposition (CBD)—the oldest and most-studied PbSe and PbS thin-film deposition method. It was used to deposit PbS in 1910.[176] The basis of CBD is a precipitation reaction between a slowly-produced anion (S^{2-} or Se^{2-}) and a complexed metal cation. The commonly used precursors are lead salts, $Pb(CH_3COO)_2$ or $Pb(NO_3)_2$, thiourea [$(NH_2)_2CS$] for PbS, and selenourea [$(NH_2)_2CSe$] for PbSe, all in alkaline solutions. Lead may be complexed with citrate, ammonia, triethanolnamine, or with selenosulfate itself. Most often, however, the deposition is carried out in a highly alkaline solution where OH^- acts as the complexing agent for Pb^{2+}.

In CBD the film is formed when the product of the concentrations of the free ions is larger than the solubility product of the compound. Thus, CBD demands very strict control over the reaction temperature, pH, and precursor concentrations. In addition, the thickness of the film is limited, the terminal thickness usually being 300–500 nm. Therefore, in order to get a film with a sufficient thickness (approximately 1 µm in IR detectors, for example), several successive depositions must be done. The benefit of CBD compared to gas phase techniques is that CBD is a low-cost temperature method and the substrate may be temperature-sensitive with the various shapes.

As-deposited PbS films exhibit significant photoconductivity. However, a post-deposition baking process is used to achieve final sensitization. In order to obtain high-performance detectors, lead chalcogenide films need to be sensitized by oxidation. The oxidation may be carried out by using additives in the deposition bath, by post-deposition heat treatment in the presence of oxygen, or by chemical oxidation of the film. The effect of the oxidant is to introduce sensitizing centers and additional states into the bandgap and thereby increase the lifetime of the photo-excited holes in the p-type material.

The backing process changes the initial n-type films to p-type films and optimizes performance through the manipulation of resistance. The best material is obtained using a specific level of oxygen and a specific bake time. Only a small percentage (3–9%) of oxygen influences the absorption properties and response of the detector. Temperatures ranging from 100 to 120°C and time periods from a few hours to in excess of 24 hours are commonly employed to achieve final detector performance optimized for a particular application. Other impurities added to the chemical-deposition solution for PbS have a considerable effect on the photosensitivity characteristics of the films.[175] $SbCl_2$, $SbCl_3$, and As_2O_3 prolong the induction period and increase the photosensitivity by up to 10 times that of films prepared without these impurities. The increase is thought to be caused by the increased absorption of CO_2 during the prolonged induction period. This increases $PbCO_3$ formation, and thus photosensitivity. Arsine sulfide also changes the oxidation states on the surface. Moreover, it has been found that essentially the same performance characteristics can be achieved by baking in an air or a nitrogen atmosphere. Therefore, all of the constituents necessary for sensitization are contained in the raw PbS films as deposited.

The preparation of PbSe photoconductors is similar to PbS ones. The post-deposition baking process for PbSe detectors operating at 77 K is carried out at a higher temperature ($>400°C$) in an oxygen atmosphere. However, for detectors to be used at ambient and/or intermediate temperatures, the oxygen or air bake is immediately followed by baking in a halogen gas atmosphere at temperatures in the range of 300 to 400°C.[177] A variety of materials can be used as substrates, but the best detector performance is achieved using single-crystal quartz material. PbSe detectors are often matched with Si to obtain higher collection efficiency.

Photoconductors also have been fabricated from epitaxial layers without backing that resulted in devices with uniform sensitivity, uniform response time, and no aging effects. However, these devices do not offset the increased difficulty and cost of fabrication.

References

1. A. Rogalski, K. Adamiec, and J. Rutkowski, *Narrow-Gap Semiconductor Photodiodes*, SPIE Press, Bellingham, WA (2000).
2. P. Capper (Ed.), *Properties of Narrow Gap Cadmium-based Compounds*, EMIS Data Reviews Series No. 10, INSPEC, IEE, London (1994).
3. P. Capper (Ed.), *Narrow-Gap II-VI Compounds for Optoelectronic and Electromagnetic Applications*, Chapman & Hall, London (1997).
4. A. Rogalski (Ed.), *Infrared Photon Detectors*, SPIE Press, Bellingham, WA (1995).
5. P. W. Kruse, "The emergence of $Hg_{1-x}Cd_xTe$ as a modern infrared sensitive material," *Semiconductors and Semimetals*, Vol. 18, R. K. Willardson and A. C. Beer (Eds.), Academic Press, New York, 1–20 (1981).
6. W. F. H. Micklethweite, "The crystal growth of mercury cadmium telluride," *Semiconductors and Semimetals*, Vol. 18, R. K. Willardson and A. C. Beer (Eds.), Academic Press, New York, 48–119 (1981).
7. R. Dornhaus, G. Nimtz, and B. Schlicht, *Narrow-Gap Semiconductors*, Springer Verlag, Berlin (1983).
8. D. Long and J. L. Schmit, "Mercury-cadmium telluride and closely related alloys," *Semiconductors and Semimetals*, Vol. 5, R. K. Willardson and A. C. Beer (Eds.), Academic Press, New York, 175–255 (1970).

9. D. G. Seiler, J. R. Lowney, C. L. Litter, and M. R. Loloee, "Energy gap versus alloy composition and temperature in $Hg_{1-x}Cd_xTe$ by two phonon magnetoabsorption techniques," *J. Vac. Sci. Technol. A* **8**, 1237–1244 (1990).

10. J. R. Lowney, D. G. Seiler, C. L. Littler, and I. T. Yoon, "Intrinsic carrier concentration of narrow gap mercury cadmium telluride," *J. Appl. Phys.* **71**, 1253–1258 (1992).

11. M. H. Weiler, "Magnetooptical properties of $Hg_{1-x}Cd_xTe$ alloys," *Semiconductors and Semimetals*, Vol. 16, R. K. Willardson and A. C. Beer (Eds.), Academic Press, New York, 119–191 (1981).

12. J. Brice and P. Capper (Eds.), *Properties of Mercury Cadmium Telluride*, EMIS Data Reviews Series No. 3, INSPEC, IEE, London (1987).

13. J. R. Meyer, C. A. Hoffman, F. J. Bartoli, D. A. Arnold, S. Sivanathan and J. P. Faurie, "Methods for magnetotransport characterization of IR detector materials," *Semicon. Sci. Technol.* **8**, 805–823 (1991).

14. J. P. Rosbeck, R. E. Star, S. L. Price, and K. J. Riley, "Background and temperature dependent current-voltage characteristics of $Hg_{1-x}Cd_xTe$ photodiodes," *J. Appl. Phys.* **53**, 6430–6440 (1982).

15. W. M. Higgins, G. N. Pultz, R. G. Roy, and R. A. Lancaster, "Standard relationships in the properties of $Hg_{1-x}Cd_xTe$," *J. Vac. Sci. Technol. A* **7**, 271–275 (1989).

16. J. Baars, D. Brink, and J. Ziegler, "Determination of acceptor densities in p-type by thermoelectric measurements," *J. Vac. Sci. Technol. B* **9**, 1709–1715 (1991).

17. P. Capper, "A review of impurity behavior in bulk and epitaxial $Hg_{1-x}Cd_xTe$," *J. Vac. Sci. Technol. B* **9**, 1667–1686 (1991).

18. H. R. Vydyanath, "Mechanisms of incorporation of donor and acceptor dopants in $Hg_{1-x}Cd_xTe$ alloys," *J. Vac. Sci. Technol. B* **9**, 1716–1723 (1991).

19. D. Shaw, "Diffusion," in *Properties of Narrow Gap Cadmium-based Compounds*, EMIS Data Reviews Series No. 10, Chapter 4A, P. Capper (Ed.), INSPEC, IEE, London, 103–142 (1994).

20. H. R. Vydyanath, "Incorporation of dopants and native defects in bulk $Hg_{1-x}Cd_xTe$ crystals and epitaxial layers," *J. Crystal Growth* **161**, 64–72 (1996).

21. H. R. Vydyanath, V. Nathan, L. S. Becker, and G. Chambers, "Materials and process issues in the fabrication of high-performance HgCdTe infrared detectors," *Proc. SPIE* **3629**, 81–87 (1999).

22. P. Capper, "Intrinsic and extrinsic doping," in *Narrow-Gap II-VI Compounds for Optoelectronic and Electromagnetic Applications*, P. Capper (Ed.), Chapman & Hall, London, 211–237 (1997).

23. Y. Marfaing, "Point defects in narrow gap II-VI compounds," in *Narrow-Gap II-VI Compounds for Optoelectronic and Electromagnetic Applications*, P. Capper (Ed.), Chapman & Hall, London, 238–267 (1997).

24. S. Sher, M. A. Berding, M. van Schlifgaarde, and An-Ban Chen, "HgCdTe status review with emphasis on correlations, native defects and diffusion," *Semiconductor Science Technology* **6**, C59–C70 (1991).

25. M. A. Berding, M. van Schilfgaarde, and A. Sher, "$Hg_{0.8}Cd_{0.2}Te$ native defect: densities and dopant properties," *J. Electron. Mater.* **22**, 1005–1010 (1993).

26. J. L. Melendez and C. R. Helms, "Process modeling of point defect effects in $Hg_{1-x}Cd_xTe$," *J. Electron. Mater.* **22**, 999–1004 (1993).

27. S. Holander-Gleixner, H. G. Robinson, and C. R. Helms, "Simulation of HgTe/CdTe interdiffusion using fundamental point defect mechanisms," *J. Electron. Mater.* **27**, 672–679 (1998).

28. P. Capper, C. D. Maxey, C. L. Jones, J. E. Gower, E. S. O'Keefe, and D. Shaw, "Low temperature thermal annealing effects in bulk and epitaxial $Cd_xHg_{1-x}Te$," *J. Electron. Mater.* **28**, 637–648 (1999).

29. M. A. Berding "Equilibrium properties of indium and iodine in LWIR HgCdTe," *J. Electron. Mater.* **29**, 664–668 (2000).

30. D. Chandra, H. F. Schaake, J. H. Tregilgas, F. Aqariden, M. A. Kinch, and A. J. Syllaios, "Vacancies in $Hg_{1-x}Cd_xTe$," *J. Electron. Mater.* **29**, 729–731 (2000).

31. H. Wiedemayer and Y. G. Sha, "The direct determination of the vacancy concentration and p-T phase diagram of $Hg_{0.8}Cd_{0.2}Te$ and $Hg_{0.6}Cd_{0.4}Te$ by dynamic mass-loss measurements," *J. Electron. Mater.* **19**, 761–771 (1990).

32. H. F. Schaake, J. H. Tregilgas, J. D. Beck, M. A. Kinch, and B. E. Gnade, "The effect of low temperature annealing on defects, impurities and electrical properties," *J. Vac. Sci. Technol. A* **3**, 143–149 (1985).

33. D. A. Stevenson and M. F. S. Tang, "Diffusion mechanisms in mercury cadmium telluride," *J. Vac. Sci. Technol. B* **9**, 1615–1624 (1991).

34. K. Yasuda, Y. Tomita, Y. Masuda, T. Ishiguro, Y. Kawauchi, and H. Morishita, "Growth condition of iodine-doped n^+-CdTe layers in MOVPE," *J. Electron. Mater.* **31**, 785–790 (2002).

35. P. Mitra, Y. L. Tyan, F. C. Case, R. Starr, and M. B. Reine, "Improved arsenic doping in metalorganic chemical vapor deposition of HgCdTe and *in situ* growth of high performance long wavelength infrared photodiodes," *J. Electron. Mater.* **25**, 1328–1335 (1996).

36. A. Piotrowski, W. Gawron, K. Kłos, J. Pawluczyk, J. Piotrowski, P. Madejczyk, and A. Rogalski, "Improvements in MOCVD growth of $Hg_{1-x}Cd_xTe$ heterostructures for uncooled infrared photodetectors," *Proc. SPIE* **5957**, 59570J (2005).

37. M. Tanaka, K. Ozaki, H. Nishino, H. Ebe, and Y. Miyamoto, "Electrical properties of HgCdTe epilayers doped with silver using an $AgNO_3$ solution," *J. Electron. Mater.* **27**, 579–582 (1998).

38. S. P. Tobin, G. N. Pultz, E. E. Krueger, M. Kestigian, K. K. Wong, and P. W. Norton, "Hall effect characterization of LPE HgCdTe p/n heterojunctions," *J. Electron. Mater.* **22**, 907–914 (1993).

39. D. Chandra, M. W. Goodwin, M. C. Chen, and L. K. Magel, "Variation of arsenic diffusion coefficients in HgCdTe alloys with temperature and Hg pressure: Tuning of p on n double layer heterojunction diode properties," *J. Electron. Mater.* **24**, 599–608 (1995).

40. D. Shaw, "Diffusion in mercury cadmium telluride—An update," *J. Electron. Mater.* **24**, 587–597 (1995).

41. L. O. Bubulac, D. D. Edwall, S. J. C. Irvine, E. R. Gertner, and S. H. Shin, "p-Type doping of double layer mercury cadmium telluride for junction formation," *J. Electron. Mater.* **24**, 617–624 (1995).

42. H. Maier and J. Hesse, "Growth, properties and applications of narrow-gap semiconductors," *Crystal Growth, Properties and Applications*, H. C. Freyhardt (Ed.), Springer Verlag, Berlin, 145–219 (1980).

43. H. R. Vydyanath, "Status of Te-rich and Hg-rich liquid phase epitaxial technologies for the growth of (Hg,Cd)Te alloys," *J. Electron. Mater.* **24**, 1275–1285 (1995).

44. R. F. Brebrick, "Thermodynamic modeling of the Hg-Cd-Te and Hg-Zn-Te systems," *J. Crystal Growth* **86**, 39–48 (1988).

45. J. C. Brice, "A numerical description of the Cd-Hg-Te phase diagram," *Prog. Cryst. Growth Charact.* **13**, 39–61 (1986).

46. T. C. Yu and R. F. Brebrick, "Phase diagrams for HgCdTe," in *Properties of Narrow Gap Cadmium-based Compounds*, EMIS Datareviews Series No. 10, P. Capper (Ed.), IEE, London, 55–63 (1994).

47. W. G. Coates, P. Capper, C. L. Jones, J. J. Gosney, C. K. Ard, I. Kenworthy, and A. Clark, "Effect of ACRT rotation parameters on Bridgman grown $Hg_{1-x}Cd_xTe$ crystals," *J. Cryst. Growth* **94**, 959–966 (1989).

48. J. H. Tregilgas, "Developments in recrystallized bulk HgCdTe," *Prog. Crystal Growth and Charact.* **28**, 57–83 (1994).

49. L. Colombo, R. R. Chang, C. J. Chang, and B. A. Baird, "Growth of Hg-based alloys by the travelling heater method," *J. Vac. Sci. Technol. A* **6**, 2795–2799 (1988).

50. R. Triboulet, A. Tromson-Carli, D. Lorans, and T. Nguyen Duy, "Substrate issues for the growth of mercury cadmium telluride," *J. Electron. Mater.* **22**, 827–834 (1993).

51. E. R. Gertner, W. E. Tennant, J. D. Blackwell, and J. P. Rode, "HgCdTe on sapphire—a new approach to infrared detector arrays," *J. Cryst. Growth* **72**, 462–467 (1987).

52. L. J. Kozlowski, S. L. Johnston, W. V. McLevige, A. H. B. Vandervyck, D. E. Cooper, S. A. Cabelli, E. R. Blazejewski, K. Vural, and W. E. Tennant, "128×128 PACE 1 HgCdTe hybrid FPAs for thermoelectrically cooled applications," *Proc. SPIE* **1685**, 193–203 (1992).

53. T. J. DeLyon, J. E. Jensen, M. D. Gorwitz, C. A. Cockrum, S. M. Johnson, and G. M. Venzor, "MBE growth of HgCdTe on silicon substrates for large-area infrared focal plane arrays: A review of recent progress," *J. Electron. Mater.* **28**, 705–711 (1999).

54. J. B. Varesi, A. A. Buell, R. E. Bornfreund, W. A. Radford, J. M. Peterson, K. D. Maranowski, S. M. Johnson, and D. F. King, "Developments in the fabrication and performance of high-quality HgCdTe detectors grown on 4-in. Si substrates," *J. Electron. Mater.* **31**, 815–821 (2002).

55. M. Carmody, J. G. Pasko, D. Edwall, R. Bailey, J. Arias, S. Cabelli, J. Bajaj, L. A. Almeida, J. H. Dinan, M. Groenert, A. J. Stolz, Y. Chen, G. Brill, and N. K. Dhar, "Molecular beam epitaxy growm long wavelength infrared HgCdTe on Si detector performance," *J. Electron. Mater.* **34**, 832–838 (2005).

56. A. Rogalski, "HgCdTe infrared detector material: history, status and outlook," *Rep. Prog. Phys.* **68**, 2267–2336 (2005).

57. P. Capper, T. Tung, and L. Colombo, "Liquid phase epitaxy," in *Narrow-Gap II-VI Compounds for Optoelectronic and Electromagnetic Applications*, P. Capper (Ed.), Chapman & Hall, London, 30–70 (1997).

58. C. A. Castro, "Review of key trends in HgCdTe materials for IR focal plane arrays," *Proc. SPIE* **2021**, 2–9 (1993).

59. B. Pelliciari, "State of art of LPE HgCdTe at LIR," *J. Crystal Growth* **86**, 146–160 (1988).

60. G. Bostrup, K. L. Hess, J. Ellsworthy, D. Cooper, and R. Haines, "LPE HgCdTe on sapphire. Status and advancements," *J. Electron. Mater.* **30**, 560–565 (2001).

61. K. D. Maranowski, J. M. Peterson, S. M. Johnson, J. B. Varesi, A. C. Childs, R. E. Bornfreund, A. A. Buell, W. A. Radford, T. J. deLyon, and J. E. Jensen, "MBE growth of HgCdTe on silicon substrates for large format MWIR focal plane arrays," *J. Electron. Mater.* **30**, 619–622 (2001).

62. T. Tung, L. V. DeArmond, R. F. Herald, P. E. Herning, M. H. Kalisher, D. A. Olson, R. F. Risser, A. P. Stevens, and S. J. Tighe, "State of the art of Hg-melt LPE HgCdTe at Santa Barbara Research Center," *SPIE Proc.* **1735**, 109–131 (1992).

63. P. W. Norton, P. LoVecchio, G. N. Pultz, J. Hughes, T. Robertson, V. Lukach, and K. Wong, "Scale-up of LPE processes for flexibility in manufacturing," *SPIE Proc.* **2228**, 73–83 (1994).

64. S. P. Tobin, M. A. Hutchins, and P. W. Norton, "Composition and thickness control of thin LPE HgCdTe layers using X-ray diffraction," *J. Electron. Mater.* **29**, 781–791 (2000).

65. Z. Djuric, "Isothermal vapour-phase epitaxy of mercury-cadmium telluride (Hg,Cd)Te," *Journal of Materials Science: Materials in Electronics* **5**, 187–218 (1995).

66. J. Piotrowski, "A new method of obtaing $Cd_x Hg_{1-x}Te$ thin films," *Electron. Technology* **5**, 87–89 (1972).

67. J. Piotrowski, Z. Djurić, W. Galus, V. Jović, M. Grudzień, Z. Djinović, and Z. Nowak, "Composition and thickness control of $Cd_x Hg_{1-x}Te$ layers grown by open tube isothermal vapour phase epitaxy," *J. Crystal Growth* **83**, 122–126 (1987).

68. K. Adamiec, M. Grudzien, Z. Nowak, J. Pawluczyk, J. Piotrowski, J. Antoszewski, J. Dell, C. Musc, and L. Faraone, "Isothermal vapor phase epitaxy as a versatile technology for infrared photodetectors," *SPIE Proc.* **2999**, 34–43 (1997).

69. S. B. Lee, L. K. Magel, M. F. S. Tang, D. A. Stevenson, J. H. Tregilgas, M. W. Goodwin, and R. L. Strong, "Characterization of isothermal vapor phase epitaxial (HgCd)Te," *J. Electron. Mater.* **A8**, 1098–1102 (1990).

70. J. Piotrowski and M. Razeghi, "Improved performance of IR photodetectors with 3D gap engineering," *SPIE Proc.* **2397**, 180–192 (1995).

71. J. C. Irvine, "Recent development in MOCVD of $Hg_{1-x}Cd_xTe$," *SPIE Proc.* **1735**, 92–99 (1992).

72. C. J. Summers, B. K. Wagner, R. G. Benz, and A. Conte Matos, "Recent advances in metalorganic molecular beam epitaxy of HgCdTe," *SPIE Proc.* **2021**, 56–66 (1994).

73. S. J. C. Irvine, "Growth of HgCdTe by vapour phase epitaxy," *Properties of Narrow Gap Cadmium-based Compounds*, EMIS Data Reviews Series No. 10, P. Capper (Ed.), INSPEC, IEE, London, 24–29 (1994).

74. M. J. Bevan, M. C. Chen, and H. D. Shih, "High-quality p-type $Hg_{1-x}Cd_xTe$ prepared by metalorganic chemical vapor deposition," *Appl. Phys. Lett.* **67**, 3450–3452 (1996).

75. S. J. C. Irvine, "Metal-organic vapour phase epitaxy," in *Narrow-gap II-VI Compounds for Optoelectronic and Electromagnetic Applications*, P. Capper (Ed.), Chapman & Hall, London, 71–96 (1997).

76. P. Mitra, F. C. Case, and M. B. Reine, "Progress in MOVPE of HgCdTe for advanced infrared detectors," *J. Electron. Mater.* **27**, 510–520 (1998).

77. P. Mitra, F. C. Case, M. B. Reine, T. Parodos, S. P. Tobin, and P. W. Norton, "MOVPE growth of HgCdTe for high performance 3–5 μm photodiodes operating at 100–180 K," *J. Electron. Mater.* **28**, 589–595 (1999).

78. P. Mitra, F. C. Case, H. L. Glass, V. M. Speziale, J. P. Flint, S. P. Tobin, and P. W. Norton, "HgCdTe growth on (552) oriented CdZnTe by metalorganic vapor phase epitaxy," *J. Electron. Mater.* **29**, 779–784 (2001).

79. C. D. Maxey, M. U. Ahmed, C. L. Jones, R. A. Catchpole, P. Capper, N. T. Gordon, M. Houlton, and T. Ashley, "Growth of long wavelength infrared MCT emitters on conducting substrates," *J. Electron. Mater.* **30**, 723–727 (2001).

80. D. Maxey, C. L. Jones, N. E. Metcalfe, R. Catchpole, M. R. Houlton, A. M. White, N. T. Gordon, and C. T. Elliott, "Growth of fully doped $Hg_{1-x}Cd_xTe$ heterostructures using a novel iodine doping source to achieve improved device performance at elevated temperatures," *J. Electron. Mater.* **25**, 1276–1285 (1996).

81. P. Mitra, T. R. Schimert, F. C. Case, R. Starr, M. H. Weiler, M. Kestigian, and M. B. Reine, "Metalorganic chemical vapor deposition of HgCdTe for photodiode applications," *J. Electron. Mater.* **24**, 661–668 (1995).

82. P. Madejczyk, A. Piotrowski, W. Gawron, K. Kłos, J. Pawluczyk, J. Rutkowski, J. Piotrowski, and A. Rogalski, "Growth and properties of MOCVD HgCdTe epilayers on GaAs substrates," *Opto-Electr. Rev.* **13**, 239–251 (2005).

83. J. M. Arias, S. H. Shin, J. G. Pasko, S. H. Shin, L. O. Bubulac, R. E. De-Wames, and W. E. Tennant, "Molecular beam epitaxy growth and *in situ* arsenic doping of p-on-n HgCdTe heterojunctions," *J. Appl. Phys.* **69**, 2143–2148 (1991).

84. J. M. Arias, J. G. Pasko, M. Zandian, S. H. Shin, G. M. Williams, L. O. Bubulac, R. E. DeWames, and W. E. Tennant, "MBE HgCdTe heterostructure p-on-n planar infrared photodiodes," *J. Electron. Mater.* **22**, 1049–1053 (1993).

85. J. M. Arias-Cortes, "MBE of HgCdTe for electro-optical infrared applications," *II-VI Semiconductor Compounds*, M. Pain (Ed.), World Scientific Publishing, Singapore, 509–536 (1993).

86. S. Sivananthan, P. S. Wijewarnasuriya, and J. P. Faurie, "Recent progress in the doping of MBE HgCdTe," *Proc. SPIE* **2554**, 55–68 (1995).

87. O. K. Wu, T. J. deLyon, R. D. Rajavel, and J. E. Jensen, "Molecular beam epitaxy of HgCdTe," in *Narrow-Gap II-VI Compounds for Optoelectronic and Electromagnetic Applications*, P. Capper (Ed.), Chapman & Hall, London, 97–130 (1997).

88. T. S. Lee, J. Garland, C. H. Grein, M. Sumstine, A. Jandeska, Y. Selamet, and S. Sivananthan, "Correlation of arsenic incorporation and its electrical activation in MBE HgCdTe," *J. Electron. Mater.* **29**, 869–872 (2000).

89. J. B. Varesi, A. A. Buell, R. E. Bornfreund, W. A. Radford, J. M. Peterson, K. D. Maranowski, S. M. Johnson, and D. F. King, "Developments in the fabrication and performance of high-quality HgCdTe detectors grown on 4-in. Si substrates," *J. Electron. Mater.* **31**, 815–821 (2002).

90. P. S. Wijewarnasuriya, M. Zandian, J. Phillips, D. Edwall, R. E. DeWames, G. Hildebrandt, J. Bajaj, J. M. Arias, A. I. D'Souza, and F. Moore, "Advances in large-area $Hg_{1-x}Cd_xTe$ photovoltaic detectors for remote-sensing applications," *J. Electron. Mater.* **31**, 726–731 (2002).

91. P. S. Wijewarnasuriyaa and S. Sivananthan, "Arsenic incorporation in HgCdTe grown by molecular beam epitaxy," *Appl. Phys. Lett.* **72**, 1694–1696 (1998).

92. O. K. Wu, D. M. Jamba, G. S. Kamath, G. R. Chapman, S. M. Johnson, J. M. Peterson, K. Kosai, and C. A. Cockrum, "HgCdTe molecular beam epitaxy technology: A focus on material properties," *J. Electron. Mater.* **25**, 423–429 (1995).

93. M. A. Berding and S. Sher, "Arsenic incorporation during MBE growth of HgCdTe," *J. Electron. Mater.* **28**, 799–803 (1999).

94. D. Rajavel and J. J. Zinck, "Growth kinetics and properties of heteroepitaxial (Cd,Zn)Te films prepared by metalorganic molecular beam epitaxy," *J. Electron. Mater.* **22**, 803–808 (1993).

95. L. H. Zhang, S. D. Pearson, W. Tong, B. K. Wagner, J. D. Benson, and C. J. Summers, "p-Type as-doping of $Hg_{1-x}Cd_xTe$ grown by MOMBE," *J. Electron. Mater.* **27**, 600–604 (1998).

96. R. Triboulet, "Alternative small gap materials for IR detection," *Semicond. Sci. Technol.* **5**, 1073–1079 (1990).

97. A. Sher, A. B. Chen, W. E. Spicer, and C. K. Shih, "Effects influencing the structural integrity of semiconductors and their alloys," *J. Vac. Sci. Technol. A* **3**, 105–111 (1985).

98. A. Wall, C. Caprile, A. Franciosi, R. Reifenberger, and U. Debska, "New ternary semiconductors for infrared applications: $Hg_{1-x}Mn_xTe$," *J. Vac. Sci. Technol. A* **4**, 818–822 (1986).

99. K. Guergouri, R. Troboulet, A. Tromson-Carli, and Y. Marfaing, "Solution hardening and dislocation density reduction in CdTe crystals by Zn addition," *J. Crystal Growth* **86**, 61–65 (1988).

100. P. Maheswaranathan, R. J. Sladek, and U. Debska, "Elastic constants and their pressure dependences in $Cd_{1-x}Mn_xTe$ with $0 \leq x \leq 0.52$ and in $Cd_{0.52}Zn_{0.48}Te$," *Phys. Rev. B* **31**, 5212–5216 (1985).

101. A. Rogalski, "$Hg_{1-x}Zn_xTe$ as a potential infrared detector material," *Prog. Quant. Electr.* **13**, 299–253 (1989).

102. A. Rogalski, "$Hg_{1-x}Mn_xTe$ as a new infrared detector material," *Infrared Phys.* **31**, 117–166 (1991).

103. A. Rogalski, *New Ternary Alloy Systems for Infrared Detectors*, SPIE Press, Bellingham, WA (1994).

104. R. T. Dalves and B. Lewis, "Zinc blende type HgTe-MnTe solid solutions—I," *J. Phys. Chem. Sol.* **24**, 549–556 (1963).

105. P. Becla, J. C. Han, and S. Matakef, "Application of strong vertical magnetic fields to growth of II–VI pseudo-binary alloys: HgMnTe," *J. Cryst. Growth* **121**, 394–398 (1992).

106. O. A. Bodnaruk, I. N. Gorbatiuk, V. I. Kalenik, O. D. Pustylnik, I. M. Rarenko, and B. P. Schafraniuk, "Crystalline structure and electro-physical parameters of $Hg_{1-x}Mn_xTe$ crystals," *Nieorganicheskie Materialy* **28**, 335–339 (1992) (in Russian).

107. R. Triboulet, "(Hg,Zn)Te: a new material for IR detection," *J. Cryst. Growth* **86**, 79–86 (1988).

108. P. Gille, U. Rössner, N. Puhlmann, H. Niebsch, and T. Piotrowski, "Growth of $Hg_{1-x}Mn_xTe$ crystals by the travelling heater method," *Semicond. Sci. Technol.* **10**, 353–357 (1995).

109. S. Takeyama and S. Narita, "New techniques for growing highly-homogeneous quaternary $Hg_{1-x}Cd_xMn_yTe$ single crystals," *Jap. J. Appl. Phys.* **24**, 1270–1273 (1985).

110. T. Uchino and K. Takita, "Liquid phase epitaxial growth of $Hg_{1-x-y}Cn_xZn_yTe$ and $Hg_{1-x}Cd_xMn_yTe$ from Hg-rich solutions," *J. Vac. Sci. Technol. A* **14**, 2871–2874 (1996).

111. A. B. Horsfall, S. Oktik, I. Terry, and A. W. Brinkman, "Electrical measurements of $Hg_{1-x}Mn_xTe$ films grown by metalorganic vapour phase epitaxy," *J. Cryst. Growth* **159**, 1085–1089 (1996).

112. R. Triboulet, A. Lasbley, B. Toulouse, and R. Granger, "Growth and characterization of bulk HgZnTe crystals," *J. Cryst. Growth* **76**, 695–700 (1986).

113. S. Rolland, K. Karrari, R. Granger, and R. Triboulet, "P-to-n conversion in $Hg_{1-x}Zn_xTe$," *Semicond. Sci. Technol.* **14**, 335–340 (1999).

114. M. A. Berding, S. Krishnamurthy, A. Sher, and A. B. Chen, "Electronic and transport properties of HgCdTe and HgZnTe," *J. Vac. Sci. Technol. A* **5**, 3014–3018 (1987).

115. R. Granger, A. Lasbley, S. Rolland, C. M. Pelletier, and R. Triboulet, "Carrier concentration and transport in $Hg_{1-x}Zn_xTe$ for x near 0.15," *J. Cryst. Growth* **86**, 682–688 (1988).

116. W. Abdelhakiem, J. D. Patterson, and S. L. Lehoczky, "A comparison between electron mobility in n-type $Hg_{1-x}Cd_xTe$ and $Hg_{1-x}Zn_xTe$," *Materials Letters* **11**, 47–51 (1991).

117. R. E. Kremer, Y. Tang, and F. G. Moore, "Thermal annealing of narrow-gap HgTe-based alloys," *J. Cryst. Growth* **86**, 797–803 (1988).

118. P. I. Baranski, A. E. Bielaiev, O. A. Bodnaruk, I. N. Gorbatiuk, S. M. Kimirenko, I. M. Rarenko, and N. V. Shevchenko, "Transport properties and recombination mechanisms in $Hg_{1-x}Mn_xTe$ alloys ($x \sim 0.1$)," *Fizyka i Technika Poluprovodnikov* **24**, 1490–1493 (1990).

119. M. M. Trifonova, N. S. Baryshev, and M. P. Mezenceva, "Electrical properties of n-type $Hg_{1-x}Mn_xTe$ alloys," *Fizyka i Technika Poluprovodnikov* **25**, 1014–1017 (1991).

120. W. A. Gobba, J. D. Patterson, and S. L. Lehoczky, "A comparison between electron mobilities in $Hg_{1-x}Mn_xTe$ and $Hg_{1-x}Cd_xTe$," *Infrared Physics* **34**, 311–321 (1993).

121. Y. Sha, C. Su, and S. L. Lehoczky, "Intrinsic carrier concentration and electron effective mass in $Hg_{1-x}Zn_xTe$," *J. Appl. Phys.* **81**, 2245–2249 (1997).

122. K. Jóźwikowski and A. Rogalski, "Intrinsic carrier concentrations and effective masses in the potential infrared detector material, $Hg_{1-x}Zn_xTe$," *Infrared Physics* **28**, 101–107 (1988).

123. C. Wu, D. Chu, C. Sun, and T. Yang, "Infrared spectroscopy of $Hg_{1-x}Zn_xTe$ alloys," *Jap. J. Appl. Phys.* **34**, 4687–4693 (1995).

124. L. Bürkle and F. Fuchs, "InAs/(GaIn)Sb superlattices: a promising material system for infrared detection," in *Handbook of Infrared Detection and Technologies*, M. Henini and M. Razeghi (Eds.), Elsevier, Oxford, 159–189 (2002).

125. G. J. Brown, F. Szmulowicz, K. Mahalingam, S. Houston, Y. Wei, A. Gon, and M. Razeghi, "Recent advances in InAs/GaSb superlattices for very long wavelength infrared detection," *Proc. SPIE* **4999**, 457–466 (2003).

126. D. L. Smith and C. Mailhiot, "Proposal for strained type II superlattice infrared detectors," *J. Appl. Phys.* **62**, 2545–2548 (1987).

127. C. Mailhiot and D. L. Smith, "Long-wavelength infrared detectors based on strained InAs–GaInSb type-II superlattices," *J. Vac. Sci. Technol. A* **7**, 445–449 (1989).

128. G. Bastard, *Wave Mechanics Applied to Semiconductor Heterostructures*, Monographies de Physiques Series, Halsted Press, New York (1988).

129. J. P. Omaggio, J. R. Meyer, R. J. Wagner, C. A. Hoffman, M. J. Yang, D. H. Chow, and R. H. Miles, "Determination of band gap and effective masses in InAs/Ga$_{1-x}$In$_x$Sb superlattices," *Appl. Phys. Lett.* **61**, 207–209 (1992).

130. C. A. Hoffman, J. R. Meyer, E. R. Youngdale, F. J. Bartoli, R. H. Miles, and L. R. Ram-Mohan, "Electron transport in InAs/Ga$_{1-x}$In$_x$Sb superlattices," *Solid State Electron.* **37**, 1203–1206 (1994).

131. C. H. Grein, P. M. Young, and H. Ehrenreich, "Minority carrier lifetimes in ideal InGaSb/InAs superlattice," *Appl. Phys. Lett.* **61**, 2905–2907 (1992).

132. C. H. Grein, P. M. Young, M. E. Flatté, and H. Ehrenreich, "Long wavelength InAs/InGaSb infrared detectors: Optimization of carrier lifetimes," *J. Appl. Phys.* **78**, 7143–7152 (1995).

133. E. R. Youngdale, J. R. Meyer, C. A. Hoffman, F. J. Bartoli, C. H. Grein, P. M. Young, H. Ehrenreich, R. H. Miles, and D. H. Chow, "Auger lifetime enhancement in InAs–Ga$_{1-x}$In$_x$Sb superlattices," *Appl. Phys. Lett.* **64**, 3160–3162 (1994).

134. O. K. Yang, C. Pfahler, J. Schmitz, W. Pletschen, and F. Fuchs, "Trap centers and minority carrier lifetimes in InAs/GaInSb superlattice long wavelength photodetectors," *Proc. SPIE* **4999**, 448–456 (2003).

135. J. C. Woolley and J. Warner, "Preparation of InAs-InSb alloys," *J. Electrochem. Soc.* **111**, 1142–1145 (1964).

136. J. C. Woolley and J. Warner, "Optical energy-gap variation in InAs-InSb alloys," *Can. J. Phys.* **42**, 1879–1885 (1964).

137. H. H. Wieder and A. R. Clawson, "Photo-electronic properties of InAs$_{0.07}$Sb$_{0.93}$ films," *Thin Solid Films* **15**, 217–221 (1973).

138. J. D. Kim, S. Kim, D. Wu, J. Wojkowski, J. Xu, J. Piotrowski, E. Bigan, and M. Razeghi, "8–13 μm InAsSb heterojunction photodiode operating at near room temperatures," *Appl. Phys. Lett.* **67**, 2645–2647 (1995).

139. J. D. Kim, D. Wu, J. Wojkowski, J. Piotrowski, J. Xu, and M. Razeghi, "Long-wavelength InAsSb photoconductors operated at near room temperatures (200–300 K)," *Appl. Phys. Lett.* **68**, 99–101 (1996).

140. G. B. Stringfellow and P. R. Greene, "Liquid phase epitaxial growth of InAs$_{1-x}$Sb$_x$," *J. Electrochem. Soc.* **118**, 805–810 (1971).

141. J. D. Kim and M. Razeghi, "Investigation of InAsSb infrared photodetectors for near-room temperature operation," *Opto-Electr. Rev.* **6**, 217–230 (1998).

142. E. Michel and M. Razeghi, "Recent advances in Sb-based materials for un-cooled infrared photodetectors," *Opto-Electr. Rev.* **6**, 11–23 (1998).

143. J. C. Woolley and B. A. Smith, "Solid solution in III-V compounds," *Proc. Phys. Soc.* **72**, 214–223 (1958).

144. M. J. Aubin and J. C. Woolley, "Electron scattering in InAsSb alloys," *Can. J. Phys.* **46**, 1191–1198 (1968).

145. E. H. Van Tongerloo and J. C. Woolley, "Free-carrier Faraday rotation in InAs$_{1-x}$Sb$_x$ alloys," *Can. J. Phys.* **46**, 1199–1206 (1968).

146. A. Rogalski, "InAs$_{1-x}$Sb$_x$ infrared detectors," *Prog. Quant. Electron.* **13**, 191– 231 (1989).

147. G. C. Osbourn, "InAsSb strained-layer superlattices for long wavelength detector," *J. Vac. Sci. Technol. B* **2**, 176–178 (1984).

148. Y. H. Zhang, A. Lew, E. Yu, and Y. Chen, "Microstructural properties of InAs/InAs$_x$Sb$_{1-x}$ ordered alloys grown by modulated molecular beam epitaxy," *J. Crystal Growth* **175/176**, 833–837 (1997).

149. M. Van Schilfgaarde, A. Sher, and A. B. Chen, "InTlSb: An infrared detector material?" *Appl. Phys. Lett.* **62**, 1857–1859 (1993).

150. A. B. Chen, M. Van Schilfgaarde, and A. Sher, "Comparison of In$_{1-x}$Tl$_x$Sb and Hg$_{1-x}$Cd$_x$Te as long wavelength infrared materials," *J. Electron. Mater.* **22**, 843–846 (1993).

151. H. Asahi, "Tl-based III-V alloy semiconductors," in *Infrared Detectors and Emitters: Materials and Devices*, P. Capper and C. T. Elliott (Eds.), Kluwer Academic Publishers, Boston, 233–349 (2001).

152. M. Van Schilfgaarde, A. B. Chen, S. Krishnamurthy, and A. Sher, "InTlP— a proposed infrared detector material," *Appl. Phys. Lett.* **65**, 2714–2716 (1994).

153. M. Razeghi, "Current status and future trends of infrared detectors," *Opto-Electron. Rev.* **6**, 155–194 (1998).

154. J. J. Lee and M. Razeghi, "Novel Sb-based materials for uncooled infrared photodetector applications," *J. Cryst. Growth* **221**, 444–449 (2000).

155. J. L. Lee and M. Razeghi, "Exploration of InSbBi for uncooled long-wavelength infrared photodetectors," *Opto-Electr. Rev.* **6**, 25–36 (1998).

156. J. N. Baillargeon, P. J. Pearah, K. Y. Cheng, G. E. Hofler, and K. C. Hsieh, "Growth and luminescence properties of GaPN," *J. Vac. Sci. Technol. B* **10**, 829–831 (1992).

157. M. Weyers, M. Sato, and H. Ando, "Red shift of photoluminescence and absorption in dilute GaAsN alloy layers," *Jap. J. Appl. Phys.* **31**(7A), L853–L855 (1992).

158. T. Ashley, T. M. Burke, G. J. Pryce, A. R. Adams, A. Andreev, B. N. Murdin, E. P. O'Reilly, and C. R. Pidgeon, "InSb$_{1-x}$N$_x$ growth and devices," *Solid-State Electronics* **47**, 387–394 (2003).

159. B. N. Murdin, M. Kamal-Saadi, A. Lindsay, E. P. O'Reilly, A. R. Adams, G. J. Nott, J. G. Crowder, C. R. Pidgeon, I. V. Bradley, J. P. Wells, T. Burke, A. D. Johnson, and T. Ashley, "Auger recombination in long-wavelength infrared InN$_x$Sb$_{1-x}$ alloys," *Appl. Phys. Lett.* **78**, 1568–1570 (2001).

160. R. Dalven, "A review of the semiconductor properties of PbTe, PbSe, PbS and PbO," *Infrared Phys.* **9**, 141–184 (1969).

161. Yu. I. Ravich, B. A. Efimova, and I. A. Smirnov, *Semiconducting Lead Chalcogenides*, Plenum Press, New York (1970).

162. J. Melngailis and T. C. Harman, "Single-crystal lead-tin chalcogenides," *Semiconductors and Semimetals*, Vol. 5, R. K. Willardson and A. C. Beer (Eds.), Academic Press, New York, 111–174 (1970).

163. T. C. Harman and J. Melngailis, "Narrow gap semiconductors," *Applied Solid State Science*, Vol. 4, R. Wolfe (Ed.), Academic Press, New York, 1–94 (1974).

164. Rogalski and J. Piotrowski, "Intrinsic infrared detectors," *Prog. Quant. Electr.* **12**, 87–289 (1988).

165. H. Preier, "Recent advances in lead-chalcogenide diode lasers," *Appl. Phys.* **10**, 189–206 (1979).

166. S. G. Parker and R. E. Johnson, "Preparation and properties of (Pb,Sn)Te," in *Preparation and Properties of Solid State Material*, Vol. 6, W. R. Wilcox (Ed.), Marcel Dekker, Inc., New York, 1–65 (1981).

167. D. L. Partin, "Molecular-beam epitaxy of IV–VI compound heterojunctions and superlattices," in *Semiconductors and Semimetals*, Vol. 33, R. K. Willardson and A. C. Beer (Eds.), Academic Press, Boston, 311–336 (1991).

168. D. Genzow, A. G. Mironow, and O. Ziep, "On the interband absorption in lead chalcogenides," *Phys. Stat. Sol.(b)* **90**, 535–542 (1978)

169. O. Ziep and D. Genzow, "Calculation of the interband absorption in lead chalcogenides using a multiband model," *Phys. Stat. Sol. (b)* **96**, 359–368 (1979).

170. W. W. Anderson, "Absorption constant of $Pb_{1-x}Sn_xTe$ and $Hg_{1-x}Cd_xTe$ alloys," *Infrared Phys.* **20**, 363–372 (1980).

171. D. E. Bode, T. H. Johnson, and B. N. McLean, "Lead selenide detectors for intermediate temperature operation," *Appl. Opt.* **4**, 327–331 (1965).

172. J. N. Humphrey, "Optimization of lead sulfide infrared detectors under diverse operating conditions," *Appl. Opt.* **4**, 665–675 (1965).

173. T. H. Johnson, H. T. Cozine, and B. N. McLean, "Lead selenide detectors for ambient temperature operation," *Appl. Opt.* **4**, 693–696 (1965).

174. D. E. Bode, "Lead salt detectors," *Physics of Thin Films*, Vol. 3, G. Hass and R. E. Thun (Eds.), Academic Press, New York, 275–301 (1966).

175. T. S. Moss, G. J. Burrel, and B. Ellis, *Semiconductor Optoelectronics*, Butterworths, London (1973).

176. R. H. Harris, "Lead-salt detectors," *Laser Focus/Electro-Optics*, 87–96 (Dec. 1983).

177. T. H. Johnson, "Lead salt detectors and arrays: PbS and PbSe," *Proc. SPIE* **443**, 60–94 (1984).

Chapter 4

Intrinsic Photodetectors

As indicated in Chapter 2 [see Eq. (2.18)], the detectivity of an IR photodetector of any type can be expressed as

$$D^* = \frac{\lambda}{2^{1/2}hc}\left(\frac{\alpha}{G+R}\right)^{1/2} F(\alpha t),$$

(4.1)

where

$$F(\alpha t) = \frac{1 - e^{-\alpha t}}{(\alpha t)^{1/2}},$$

(4.2)

α is the absorption coefficient, G and R are the generation and recombination rates, and t is the detector's thickness.

This chapter considers the general theory of intrinsic IR photodetectors, including decisive generation-recombination mechanisms in semiconductor materials at near room temperature.

4.1 Optical Absorption

Optical properties of narrow-gap semiconductors have been widely investigated.[1] The absorption at the band edge is essential to determine the performance of IR photodetectors. However, there still appears to be considerable disagreement among the reported results concerning absorption coefficients at wavelengths close to those corresponding with the bandgap. In most compound semiconductors, the band structure closely resembles the parabolic energy versus the momentum dispersion relation. The optical absorption coefficient would then have a square-root dependence on energy that follows the electronic density of states, often referred to as the Kane model.[2] The above bandgap absorption coefficient can be calculated for InSb-like band structure semiconductors such as $Hg_{1-x}Cd_xTe$, including the Moss-Burstein shift effect. Corresponding expressions were derived by Anderson.[3] Beattie and White proposed an analytic approximation with a wide

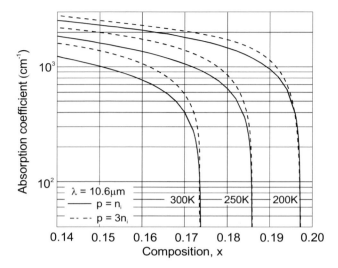

Figure 4.1 Calculated absorption coefficient of HgCdTe 10.6 μm as a function of composition, temperature, and doping.

range of applicability for band-to-band radiative transition rates in direct, narrow-bandgap semiconductors.[4]

The Kane region edges are steep as one would expect with a direct-gap semiconductor. In general, the absorption coefficient depends on doping due to changes to the Fermi level as a result of the Burstein-Moss effect. This is especially true for narrow-bandgap material at high temperatures. To give an example, Fig. 4.1 shows the calculated absorption coefficient of $Hg_{1-x}Cd_xTe$ at 10.6 μm within the Kane model as a function of composition, temperature, and doping. The p-type doping increases absorption due to the reduced bandfilling effect in the conductance band. The increase of absorption with doping levels used for device fabrication of uncooled detectors is quite significant (a factor of ≈ 2 for $x \approx 0.17$ and $N_a = 2 \times 10^{17}$ cm^{-3}).

While the measured absorption for $h\nu > E_g$ is in good agreement with the Kane model, the long-wavelength absorption edge in $Hg_{1-x}Cd_xTe$ and closely related semiconductor alloys ($Hg_{1-x}Zn_xTe$, $Hg_{1-x}Mn_xTe$) exhibit an exponential tail extending at energies lower than the bandgap. The tail has been attributed to a composition-induced disorder. The tailing is additionally increased by native point defects, impurities, nonuniform composition, and other imperfections of the crystals.

Near the band edge, a band-tail absorption typically follows an exponential dependence versus energy, commonly referred to as an Urbach tail:[5, 6]

$$\alpha = \alpha_o \exp\left[\frac{\sigma(E - E_o)}{T + T_o}\right], \tag{4.3}$$

where E_o is the transition energy between the below-bandgap and above-bandgap regions of the absorption coefficient, α_o is the corresponding coefficient at this energy, and $\sigma/(T + T_o)$ is the exponential tail slope parameter. According to Finkman and Schacham, the absorption tail obeys the rule [see (Eq. 4.3)], where α in cm^{-1}, E is in eV, T is in K, $T_o = 81.9$ (in K), $E_o = -0.3424 + 1.838x + 0.148x^2$ (in eV), $\sigma = 3.267 \times 10^4(1 + x)$ (in K/eV), and $\alpha_o = \exp(53.61x - 18.88)$. These are all fitting parameters that vary smoothly with composition.[7] The fit was performed with data at $x = 0.215$ and $x = 1$, and for temperatures between 80 and 300 K.

Assuming that the absorber coefficient for large energies can be expressed as

$$\alpha(h\nu) = \beta(h\nu - E_g)^{1/2}, \tag{4.4}$$

many researchers assume that this rule can be applied to HgCdTe.[6–8] For example, Schacham and Finkman used the following fitting parameter $\beta = 2.109 \times 10^5[(1 + x)/(81.9 + T)]^{1/2}$, which is a function of composition and temperature.[8] The conventional procedure used to locate the energy gap is to use the point inflection, i.e., exploit the large change in the slope of $\alpha(h\nu)$ that is expected when the band-to-band transition overtakes the weaker Urbach contribution. To overcome the difficulty in locating the onset of the band-to-band transition, the bandgap was defined as that energy value where $\alpha(h\nu) = 500$ cm^{-1}.[7] Schacham and Finkman[8] analyzed the crossover point and suggested $\alpha = 800$ cm^{-1} was a better choice. Hougen[9] analyzed absorption data of n-type LPE layers and suggested that the best formula was $\alpha = 100 + 5000x$.

Chu et al.[10] reported similar empirical formulas for the absorption coefficient at the Kane and Urbach tail regions. They received the following modified Urbach rule of the form:

$$\alpha = \alpha_o \exp\left[\frac{\delta(E - E_o)}{kT}\right], \tag{4.5}$$

where

$$\ln \alpha_o = -18.5 + 45.68x,$$

$$E_o = -0.355 + 1.77x,$$

$$\delta/kT = \frac{(\ln \alpha_g - \ln \alpha_o)}{(E_g - E_o)},$$

$$\alpha_g = -65 + 1.88T + (8694 - 10.315T)x, \text{ and}$$

$$E_g(x, T) = -0.295 + 1.87x - 0.28x^2 + 10^{-4}(6 - 14x + 3x^2)T + 0.35x^4.$$

The meaning of the parameter α_g is that $\alpha = \alpha_g$ when $E = E_g$, the absorption coefficient at the bandgap energy. When $E < E_g$, $\alpha < \alpha_g$, the absorption coefficient obeys the Urbach rule in Eq. (4.5).

Chu et al.[11] also found an empirical formula to calculate the intrinsic optical absorption coefficient at the Kane region:

$$\alpha = \alpha_g \exp[\beta(E - E_g)]^{1/2}, \qquad (4.6)$$

where the parameter β depends on the alloy composition and temperature

$$\beta(x, T) = -1 + 0.083T + (21 - 0.13T)x.$$

By expanding Eq. (4.6), one finds a linear term, $(E - E_g)^{1/2}$, that fits the square-root law between α and E proper for parabolic bands [see Eq. (4.4)].

Figure 4.2 shows the intrinsic absorption spectrum for $Hg_{1-x}Cd_x Te$ with $x = 0.170 - 0.443$ at temperatures 300 K and 77 K. It can be seen that the calculated Kane plateaus according to Ref. 12 and Eq. (4.6) link closely with the calculated Urbach absorption tail from Eq. (4.5) at the turning point α_g. Since the tail effect is not included in the Anderson model,[3] the curves calculated according to this model fall down sharply at energies adjacent to E_g. At 300 K, the line shapes derived for absorption coefficients above α_g have almost the same tendency; however, Chu et al.'s expression [Eq. (4.5)] shows a stronger agreement with the experimental data. At 77 K, the curves for the expressions of Anderson and Chu et al. are in agreement with the measurements, but discrepancies occur for the empirical parabolic rule of Sharma et al.,[12] and these deviations increase with decreasing x. The degree of band nonparabolicity increases as the temperature or x decrease, which results in an increasing discrepancy between the experimental result and the square-root law. In general, Chu et al.'s empirical rule and the Anderson model agree with experimental data for $Hg_{1-x}Cd_x Te$, with x ranging from 0.170 to 0.443 and at temperatures from 4.2 to 300 K, but the Anderson model fails to explain the absorption near E_g.[13]

More recently, it has been suggested that narrow-bandgap semiconductors such as HgCdTe more closely resemble a hyperbolic band-structure relationship with an absorption coefficient given by

$$\alpha = \frac{K\sqrt{(E - E_g + c)^2 - c^2}(E - E_g + c)}{E}, \qquad (4.7)$$

where c is the parameter defining hyperbolic curvature of the band structure, and K is the parameter defining the absolute value of the absorption coefficient.[14, 15] Recently this theoretical prediction has been confirmed by experimental measurements of optical properties of MBE HgCdTe-grown samples with uniform compositions.[16, 17] As defined in Eqs. (4.3) and (4.7), the fitting parameters for bandgap, tail, and hyperbolic regions of the absorption coefficient have been extracted by the determination of the transition point between both regions. The derivative of the absorption coefficient has a maximum between the Urbach and hyperbolic regions. Figure 4.3 shows the measured exponential-slope parameter values $\sigma/(T + T_o)$ versus temperature and compares them to values given by Finkman

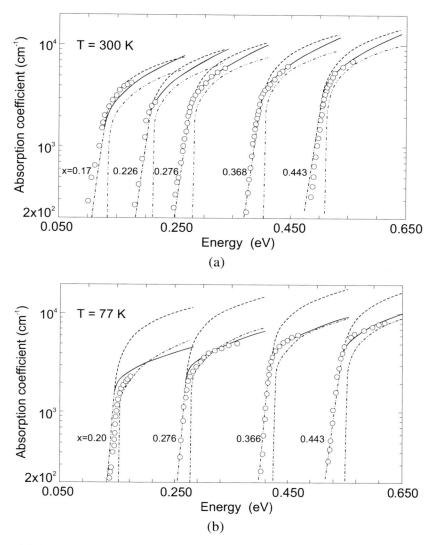

Figure 4.2 Intrinsic absorption spectrum of $Hg_{1-x}Cd_xTe$ samples with $x = 0.170 - 0.443$ at (a) 300 K, and (b) 77 K. Symbols indicate experimental data from Refs. 10 and 11: the dash-double-dotted curves are according to Anderson's model; medium dashed curves are from Sharma et al.[12]; solid curves are from Eq. (4.6) of Chu et al.; and dash-dotted lines below E_g are from Eq. (4.5). (Reprinted from Ref. 13 with permission from Elsevier.)

and Schacham for the arbitrarily chosen composition of $x = 0.3$.[7] This choice of composition does not have a significant effect on the values obtained, where the values given by Finkman and Schacham have small compositional dependence in the region of interest ($0.2 < x < 0.6$), and where the parameter $\sigma/(T + T_o)$ is proportional to $(1 + x)^3$. This parameter shows no clear correlation with composition where there is significant scatter in the data at cryogenic temperatures. The trend of decreasing values with increasing temperature is in agreement with an increase

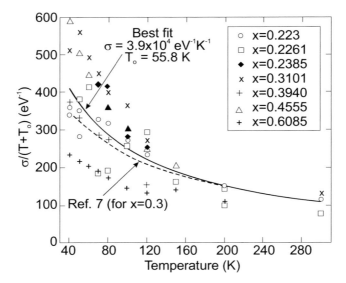

Figure 4.3 Band-tail parameter $\sigma/(T + T_o)$ versus temperature for different compositions and a proposed model based on the best overall fit along with values from Ref. 7 for $x = 0.3$. (Reprinted from Ref. 17 with permission from TMS.)

in thermally excited absorption processes, where the values obtained at lower temperatures are more indicative of the quality of the layers grown.

It should be noted that the expressions discussed above do not take into account the influence of doping on the absorption coefficient, and as a result, these expressions are not very useful in modeling the long-wavelength uncooled devices.

The standard approximations used at Vigo System S.A. are Kane calculations for large energies that take into account the doping-dependent Burstein-Moss effect and the Urbach-type expression for longer wavelengths with a 500 cm^{-1} crossover point.[18] In this case, the crossover point depends on doping.

$Hg_{1-x}Cd_x$Te and closely related alloys exhibit significant absorption, seen below the absorption edges, which can be related to intraband transitions in both the conduction and valence bands as well as the intervalence band transitions.[19] This absorption does not contribute to the optical generation of charge carriers.

4.2 Thermal Generation-Recombination Processes

The generation processes that compete against the recombination processes directly affect the performance of photodetectors. These processes set up a steady-state concentration of carriers in semiconductors that are subjected to thermal and optical excitation. The generation processes also frequently determine the kinetics of photogenerated signals.

The generation-recombination processes in semiconductors are widely discussed in the literature (see, for example, Refs. 20–22). Here we have reproduces only some dependencies directly related to photodetector performance. Assuming bulk processes only, there are three main thermal generation-recombination processes to be considered in narrow-bandgap semiconductors: Shockley-Read, radiative, and Auger.

4.2.1 Shockley-Read processes

The Shockley-Read (SR) mechanism is not an intrinsic and fundamental process because it occurs at levels in the forbidden energy gap. The reported positions of SR centers for both n- and p-type materials range anywhere from near the valence to near the conduction band.

The SR mechanism is probably responsible for lifetimes in lightly doped n- and p-type $Hg_{1-x}Cd_x Te$. The possible factors are SR centers associated with native defects and residual impurities. In n-type material ($x = 0.20$–0.21, 80 K) with carrier concentrations less than 10^{15} cm^{-3}, the lifetimes exhibit a broad range of values (0.4–8 μs) for material prepared by various techniques.[23] Dislocations may also influence the recombination time for dislocation densities $>5 \times 10^5$ cm^{-2}.[24–26]

In p-type HgCdTe, the SR mechanism is usually blamed for reducing the lifetime with decreasing temperature. The steady-state, low-temperature photoconductive lifetimes are usually much shorter than the transient lifetimes. The low-temperature lifetimes exhibit very different temperature dependencies with a broad range of values over three orders of magnitude, from 1 ns to 1 μs (p $\approx 10^{16}$ cm^{-3}, $x \approx 0.2$, T ≈ 77 K, vacancy doping).[23, 27] This is due to many factors that may affect the measured lifetime, including inhomogeneities, inclusions, and surface and contact phenomena. The highest lifetime was measured in high-quality undoped and extrinsically doped materials grown by low-temperature epitaxial techniques from Hg-rich LPE[28] and MOCVD.[29–31] Typically, Cu- or Au-doped materials exhibit lifetimes one order of magnitude larger than vacancy-doped materials of the same hole concentration.[27] It is believed that the increase of lifetime in impurity-doped $Hg_{1-x}Cd_x Te$ arises from a reduction of SR centers. This may be due to the low-temperature growth of doped layers or due to low-temperature annealing of doped samples.

The origin of the SR centers in vacancy-doped p-type material is not clear at present. These centers do not seem to be the vacancies themselves and thus may be removable.[32] Vacancy-doped material with the same carrier concentration but created under different annealing temperatures may produce different lifetimes. One possible candidate for recombination centers is Hg interstitials.[33] Vacancy-doped $Hg_{1-x}Cd_x Te$ exhibits SR recombination center densities roughly proportional to the vacancy concentration.

Measurements by DRS Technologies[34] give lifetimes values for extrinsic p-type material:

$$\tau_{ext} = 9 \times 10^9 \frac{p_1 + p}{p N_a}, \tag{4.8}$$

where

$$p_1 = N_v \exp\left[\frac{q(E_r - E_g)}{kT}\right], \qquad (4.9)$$

and E_r is the SR center energy relative to the conduction band. Experimentally, E_r was found to lie at the intrinsic level for As, Cu, and Au dopants, giving $p_1 = n_i$.

For vacancy-doped p-type $Hg_{1-x}Cd_xTe$,

$$\tau_{vac} = 5 \times 10^9 \frac{n_1}{pN_{vac}}, \qquad (4.10)$$

where

$$n_1 = N_c \exp\left(\frac{qE_r}{kT}\right). \qquad (4.11)$$

E_r is ≈ 30 mV from the conduction band ($x = 0.22$–0.30).

As follows from these expressions and Fig. 4.4, doping with the foreign impurities (Au, Cu, and As for p-type material) gives lifetimes significantly increased compared to native doping of the same level.

Although a considerable research effort is still necessary, the Shockley-Read process does not represent a fundamental limit to photodetector performance.

4.2.2 Internal radiative processes

Radiative generation of charge carriers is a result of the absorption of internally generated photons. The radiative recombination is an inversed process of annihilation of electron-hole pars with emission of photons. The radiative recombination

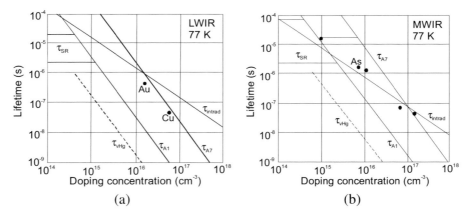

Figure 4.4 Measured lifetimes at 77 K for n- and p-type (a) LWIR, and (b) MWIR, which are compared to the theory for Auger 1, Auger 7, Shockley-Read, and internal radiative recombinations as a function of doping concentration. (Reprinted from Ref. 35 with permission from TMS.)

rates were calculated for conduction-to-heavy-hole-band and conduction-to-light-hole-band transitions using an accurate analytical form.[4]

For a long time, internal radiative processes have been considered to be the main fundamental limit to detector performance, and the performance of practical devices has been compared to that limit. The role of the radiative mechanism in IR radiation detection has been critically re-examined.[36–38] Humpreys indicated that most of the photons emitted in photodetectors as a result of radiative decay are immediately reabsorbed, and as a result the observed radiative lifetime is only a measure of how well photons can escape from the detector body.[37] Due to reabsorption, the radiative lifetime is highly extended and dependent on the semiconductor geometry. Therefore, internal combined recombination-generation processes in one detector are essentially noiseless. In contrast, the recombination act with cognate escape of a photon from the detector, or the generation of photons by thermal radiation from outside the active body of the detector, are both noise-producing processes. This may readily happen in the case of a detector array where an element may absorb photons emitted by another detector or a passive part of the structure.[38, 39] The deposition of the reflective layers (mirrors) on the back and side of the detector may significantly improve optical insulation, which prevents noisy emissions and the absorption of thermal photons.

It should be noted that internal radiative generation could be suppressed in detectors operated under reverse bias, where the electron density in the active layer is reduced to well below its equilibrium level.[40, 41]

As follows from the above considerations, the internal radiative processes—although of a fundamental nature—do not limit the ultimate performance of IR detectors.

4.2.3 Auger processes

Auger mechanisms dominate the generation and recombination processes in high-quality narrow-gap semiconductors such as $Hg_{1-x}Cd_xTe$ and InSb at near room temperatures.[42, 43] The Auger generation is essentially the electrons' impact ionization of holes in the high-energy tail of a Fermi-Dirac distribution. The band-to-band Auger mechanisms in InSb-like band structure semiconductors are classified into 10 photonless mechanisms. Two of them have the smallest threshold energies ($E_T \approx E_g$) and are denoted as Auger 1 (A1) and Auger 7 (A7). In some wider-bandgap materials (e.g., InAs and low x $InAs_{1-x}Sb_x$) in which the split-off band energy Δ is comparable to E_g, the Auger process involves a split-off band (AS process) that also may play an important role.

The A1 generation is the impact ionization by an electron, which generates an electron-hole pair, so this process involves two electrons and one heavy hole. An interesting feature is the behavior of the A1 generation and recombination with degenerate n-type doping. Because the low density of the Fermi level moves high into the conduction band with n-type doping, the concentration of minority holes is strongly reduced and the threshold energy required for the Auger transition increases. This results in suppression of A1 processes in heavy doped n-type material.

The A7 generation is the impact generation of an electron hole pair by a light hole, involving one heavy hole, one light hole, and one electron.[44-46] This process may dominate in p-type material. Heavy p-type doping has a dramatic effect on the A7 generation and recombination rates due to the much higher density of states. The corresponding Auger recombination mechanisms are inverse processes of electron-hole recombination with energy transferred to the electron or hole. Strong temperature and bandgap dependence is expected since a lowered temperature and increased bandgap strongly reduce the probability of these heat-stimulated transitions.

The net generation rate due to the A1 and A7 processes can be described as

$$G_A - R_A = \frac{n_i^2 - np}{2n_i^2}\left[\frac{n}{(1+an)\tau_{A1}^i} + \frac{p}{\tau_{A7}^i}\right], \tag{4.12}$$

where τ_{A1}^i and τ_{A7}^i are the intrinsic A1 and A7 recombination times, and n_i is the intrinsic concentration.[47] The last equation is valid for a wide range of concentrations, including degeneration, which easily occurs in n-type materials. This is expressed by the finite value of a. According to Ref. 47, $a = 5.26 \times 10^{-18}$ cm^3. Due to the shape of the valence band, the degeneracy in p-type material occurs only at very high doping levels, which is not achievable in practice.

The A1 intrinsic recombination time is equal to

$$\tau_{A1}^i = \frac{h^3\varepsilon_o^2}{2^{3/2}\pi^{1/2}q^4m_o}\frac{\varepsilon^2(1+\mu)^{1/2}(1+2\mu)\exp\{[(1+2\mu)/(1+\mu)](E_g/kT)\}}{(m_e^*/m)|F_1F_2|^2(kT/E_g)^{3/2}},$$

$$\tag{4.13}$$

where μ is the ratio of the conduction to the heavy-hole valence-band effective mass, ε_s is the static-frequency dielectric constant, and $|F_1F_2|$ are the overlap integrals of the periodic part of the electron wave functions. The overlap integrals cause the biggest uncertainly in the A1 lifetime. Values ranging from 0.1 to 0.3 have been obtained by various authors. In practice, it is taken as a constant equal to anywhere between 0.1 and 0.3, leading to changes in the lifetime by almost one order of magnitude.

The static dielectric constants for Hg$_{1-x}$Cd$_x$Te were approximated from smoothed median values:

$$\varepsilon_s = 20.5 - 15.5x + 5.7x^2.^{48} \tag{4.14}$$

The ratio of A7 and A1 intrinsic times,

$$\gamma = \frac{\tau_{A7}^i}{\tau_{A1}^i}, \tag{4.15}$$

is another term of high uncertainty. According to Casselman et al.,[44, 45] for $Hg_{1-x}Cd_xTe$ over the range $0.16 \leq x \leq 0.40$ and $50\ K \leq T \leq 300\ K$, $3 \leq \gamma \leq 6$. Direct measurements of carrier recombination show the ratio γ larger than expected from previous calculations (≈ 8 for $x \approx 0.2$ at 295 K).[49] Recently published theoretical[50, 51] and experimental[35, 51] results indicate that this ratio is even higher; the data presented in Fig. 4.4 would indicate a value of approximately 60. Since γ is higher than unity, higher recombination lifetimes are expected in p-type materials compared to n-type materials of the same doping.

Kinch delivered a simplified formula for the A1 intrinsic recombination time:

$$\tau_{A1}^i = 8.3 \times 10^{-13} E_g^{1/2} \left(\frac{q}{kT}\right)^{3/2} \exp\left(\frac{qE_g}{kT}\right), \qquad (4.16)$$

where E_g is in eV.[34]

As Eqs. (4.12) and (4.13) show, the Auger generation and recombination rates are strongly dependent on temperature via the dependence of the carrier concentration and the intrinsic time on temperature. Therefore, cooling is a natural and very effective way to suppress Auger processes.

Until recently, the n-type A1 lifetime was deemed to be well established. Krishnamurthy et al.[51] showed that the full band calculations indicate that the radiative and Auger recombination rates are much slower than those predicted by expressions used in the literature (based on the theory of Beattie and Landsberg[42]). It appears that a trap state tracking the conduction band edge with a very small activation energy can explain the lifetimes in n-doped MBE samples.

The p-type A7 lifetime has long been a subject of controversy. Beattie and White derived an analytic approximation with a wide range of applicability for Auger transitions.[52] The flat valence band model has been used to obtain a simple analytic approximation that requires just two parameters to cover a wide range of temperature and carrier Fermi levels, both degenerate and nondegenerate. Detailed calculations of the Auger lifetime in p-type HgCdTe reported by Krishnamurthy and Casselman suggest a significant deviation from the classic $\tau_{A7} \sim p^{-2}$ relation.[50] The decrease of τ_{A7} with doping is much weaker, resulting in significantly longer lifetimes in highly doped p-type low x materials (a factor of ≈ 20 for $p = 1 \times 10^{17}\ cm^{-3}$, $x = 0.226$, and $T = 77 \div 300\ K$).

4.3 Auger-Dominated Performance

4.3.1 Equilibrium devices

Let us consider the Auger-limited detectivity of photodetectors. At equilibrium, the generation and recombination rates are equal. If it is assumed that both rates contribute to noise [see Eq. (2.16)], then

$$D^* = \frac{\lambda\eta}{2hc(G_At)^{1/2}} \left(\frac{A_o}{A_e}\right)^{1/2}. \qquad (4.17)$$

If nondegenerate statistics are assumed, then

$$G_A = \frac{n}{2\tau_{A1}^1} + \frac{p}{2\tau_{A7}^1} = \frac{1}{2\tau_{A1}^1}\left(n + \frac{p}{\gamma}\right). \qquad (4.18)$$

The resulting Auger generation achieves its minimum in only extrinsic p-type materials with $p = \gamma^{1/2}n_i$ and leads to an important conclusion about optimum doping for the best performance. In practice, the required p-type doping level would be difficult to achieve for LN-cooled and short- wavelength devices. More-over, the p-type devices are more vulnerable to nonfundamental limitations (such as contacts, surface, and Shockley-Read processes) than the n-type devices. This is the reason why low-temperature and short-wavelength photodetectors are typi-cally manufactured from lightly doped n-type materials. In contrast, p-type doping is clearly advantageous for the near-room-temperature and long-wavelength pho-todetectors.

The Auger-dominated detectivity is

$$D^* = \frac{\lambda}{2^{1/2}hc}\left(\frac{A_o}{A_e}\right)^{1/2}\frac{\eta}{t^{1/2}}\left(\frac{\tau_{A1}^i}{n + p/\gamma}\right)^{1/2}. \qquad (4.19)$$

As follows from Eq. (2.23), for the optimum thickness devices,

$$D^* = 0.31k\frac{\lambda\alpha^{1/2}}{hc}\frac{(2\tau_{A1}^i)^{1/2}}{(n + p/\gamma)}. \qquad (4.20)$$

This expression can be used to determine the optimum A1/A7 detectivity as a func-tion of wavelength, material bandgap, and doping.

To estimate the wavelength and temperature dependence of D^*, let us assume a constant absorption for photons with energy equal to the bandgap. For extrinsic materials ($p = N_a$ or $n = N_d$),

$$D^* \sim \left(\tau_{A1}^i\right)^{1/2} \sim n_i^{-1} \sim \exp\left(\frac{E_g}{2kT}\right) = \exp\left(\frac{hc/2\lambda_{co}}{kT}\right). \qquad (4.21)$$

In this case the ultimate detectivity will be inversely proportional to intrinsic con-centration. This behavior should be expected at shorter wavelengths and lower tem-peratures when the intrinsic concentration is low.

For intrinsic materials and for materials doped for the minimum thermal gener-ation, where $p = \gamma^{1/2}n_i$, $n = n_i/\gamma^{1/2}$ and $n + p = 2\gamma^{-1/2}n_i$, stronger $D^* \sim n_i^{-2}$ dependence can be expected:

$$D^* \sim \frac{(\tau_{A1}^i)^{1/2}}{n_i} \sim n_i^{-2} \sim \exp\left(\frac{E_g}{kT}\right) = \exp\left(\frac{hc/\lambda_{co}}{kT}\right). \qquad (4.22)$$

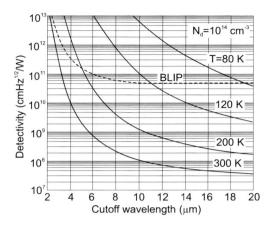

Figure 4.5 Calculated performance of Auger generation-recombination limited $Hg_{1-x}Cd_xTe$ photodetectors as a function of wavelength and temperature of operation. BLIP detectivity has been calculated for 2π FOV, $T_B = 300$ K, and $\eta = 1$.

Figure 4.5 shows the calculated detectivity of Auger generation-recombination limited $Hg_{1-x}Cd_xTe$ photodetectors as a function of wavelength and the temperature of the operation. Calculations have been performed for 10^{14} cm^{-3} doping, which is the lowest donor doping level that is achievable in a controllable manner in practice. Values as low as $\approx 1 \times 10^{13}$ cm^{-3} are at present achievable in laboratories, while values of 3×10^{14} cm^{-3} are more typical in industry. Liquid nitrogen cooling potentially makes it possible to achieve BLIP (300 K) performance over the entire 2–20 µm range. Achievable with Peltier coolers, 200 K cooling would be sufficient for BLIP operation in the middle- and short-wavelength regions (<5 µm).

Detector performance can be improved further by the use of optical immersion (Fig. 4.6). However, the theoretical detectivity limit for uncooled detectors remains approximately one or almost two orders of magnitude below D^*_{BLIP} (300 K, 2π) for $\approx$5-µm and 10-µm wavelengths, respectively. An improvement by a factor of $\approx$2 is still possible with optimum p-type doping.

4.3.2 Nonequilibrium devices

Auger generation had appeared to be a fundamental limitation to IR photodetector performance until British workers proposed a new approach to reducing the photodetector cooling requirements.[53, 54] This approach, one of the most exciting events in the field of IR photodetectors operating without cryogenic cooling, is based on the nonequilibrium mode of operation.

Their concept relies on the dependence of the Auger processes on a concentration of free carriers. The suppression of Auger processes may be achieved by decreasing the free-carrier concentration below equilibrium values. Nonequilibrium depletion of semiconductors can be applied to reduce the concentration of the majority and minority carrier concentrations. This can be achieved in some devices

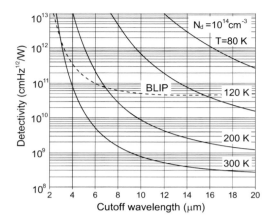

Figure 4.6 Calculated performance of optically immersed equilibrium mode Hg$_{1-x}$Cd$_x$Te photodetectors as a function of wavelength and temperature. Hyperspherical immersion and a lens refraction coefficient of $n = 2.7$ has been assumed. BLIP detectivity has been calculated for 2π FOV, $T_B = 300$ K, and $\eta = 1$.

that are based on lightly doped narrow-gap semiconductors, such as in biased low-high (l-h) doped contact structures, heterojunction contact structures, MIS structures, or structures that use the magnetoconcentration effect. Under strong depletion, the concentrations of both majority and minority carriers can be reduced below the intrinsic concentration. The majority carrier concentration saturates at the extrinsic level while the concentration of minority carriers is reduced below the extrinsic level. Therefore, the necessary condition for deep depletion is a very light doping of the semiconductor below the intrinsic concentration.

Table 4.1 explains the symbols employed to describe semiconductors used for the absorber and contact regions of a photodetector.

Table 4.1 Terminology used to describe various device component categories.

Category	Description
n	extrinsically n-doped, $N_d > n_i$, active region
p	extrinsically p-doped, $N_a > n_i$, active region
n$^-$	just extrinsic n, $n_i < N_d \lesssim 2n_i$, active region
p$^-$	just extrinsic p, $n_i < N_a \lesssim 2n_i$, active region
n$^+$	degenerated n-doped
p$^+$	degenerated p-doped
ν	near intrinsic n-type, $N_d < n_i$, active region
π	near intrinsic p-type, $N_a < n_i$, active region
N	wider bandgap than active region, extrinsically n-doped, $N_d > n_i$
P	wider bandgap than active region, extrinsically n-doped, $N_a > n_i$
N$^-$	low n-doped N material
P$^-$	low p-doped P material
N$^+$	degenerated N material
P$^+$	degenerated P material

To discuss the fundamental performance limits of Auger-suppressed devices, first consider a detector based on ν-type material. At strong depletion, $n = N_d$, and the Auger generation rate is

$$G_A = \frac{N_d}{2\tau_{A1}^i}.$$ (4.23)

As follows from Eq. (4.23), deep depletion makes the recombination rate negligible compared to the generation rate; as a result, it is readily eliminated as a noise source in depleted materials. Therefore,

$$D^* = \frac{\lambda}{hc} \frac{\eta}{t^{1/2}} \left(\frac{\tau_{A1}^i}{N_d} \right)^{1/2}.$$ (4.24)

Similarly, for π-type materials,

$$G_A = \frac{N_a}{2\gamma\tau_{A1}^i},$$ (4.25)

and

$$D^* = \frac{\lambda}{hc} \frac{\eta}{t^{1/2}} \left(\frac{\tau_{A1}^i}{N_a/\gamma} \right)^{1/2}.$$ (4.26)

Again, as in the equilibrium case, the use of π-type material is advantageous, improving the detectivity by a factor of $\gamma^{1/2}$ compared to ν-type material of the same doping ($\gamma > 1$).

Comparing the corresponding equations for equilibrium and nonequilibrium modes, we find that the use of the nonequilibrium mode of operation may reduce the Auger generation rate by a factor n_i/N_d in lightly doped material with corresponding detectivity improvement of $(2n_i/N_d)^{1/2}$. The additional gain factor of $2^{1/2}$ is due to the negligible recombination rate in the depleted semiconductor. The gain for p-type material is even larger—a factor of $[2(\gamma+1)n_i/N_a]^{1/2}$, taking into account the elimination of A1 and A7 recombinations. Additional depletion-related improvement also can be expected from increased absorption due to the reduced bandfiling effect.

The resulting improvement may be quite large, particularly for LWIR devices operating at near room temperatures as shown in Fig. 4.7. Potentially, the BLIP performance can be obtained without cooling. The BLIP limit can be achieved by the following:

- Using materials with controlled doping at a very low level ($\approx 10^{12}$ cm^{-3});
- Using extremely high-quality materials with a very low concentration of Shockley-Read centers;

- Properly designing a device that prevents thermal generation at surfaces, interfaces, and contacts; and
- Using a thermal dissipation device whose design makes it possible to achieve a state of strong depletion.

The requirements for BLIP performance ($\lambda = 10$ μm, $T = 300$ K), particularly the doping concentration, can be eased significantly by using optical immersion

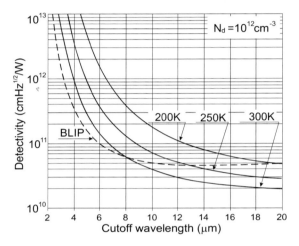

Figure 4.7 Calculated performance of Auger generation-recombination limited and operated in the nonequilibrium-mode $Hg_{1-x}Cd_xTe$ photodetectors as a function of wavelength and temperature. BLIP detectivity has been calculated for 2π FOV, $T_B = 300$ K, and $\eta = 1$.

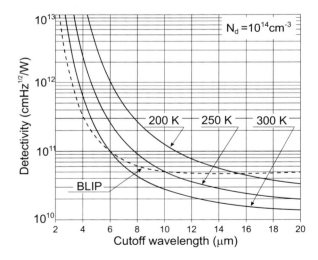

Figure 4.8 Calculated performance of optically immersed, nonequilibrium mode $Hg_{1-x}Cd_xTe$ photodetectors as a function of wavelength and temperature. Hyperspherical immersion and a lens refraction coefficient of $n = 2.7$ has been assumed. BLIP detectivity has been calculated for 2π FOV, $T_B = 300$ K, and $\eta = 1$.

(Fig. 4.8). In principle, BLIP performance can be obtained with an $\approx 10^{13}$ cm^{-3} donor and $\approx 3 \times 10^{13}$ cm^{-3} acceptor doping.

4.4 Modeling of High-Temperature Photodetectors

Traditionally IR photodetectors have been called either photoconductive or photovoltaic detectors based on the principle by which optically generated carriers can be detected as a change in voltage or current across an element.[55] The simple design of photoconductors is based on a flake of semiconductor supplied with ohmic contacts, while photovoltaic detectors are p-n homojunction devices. Dember and PEM-effect detectors are less-common photovoltaic devices that require no p-n junction.

However, recent advances in heterostructure devices such as the development of the heterojunction photoconductor and the double-layer heterostructure photodiode, and the introduction of the nonequilibrium mode of operation, make this distinction unclear. Moreover, photovoltaic structures are frequently biased, exhibiting signal due to both photovoltages at junctions and from the photoconductivity of some regions.

An optimized photodetector (see Fig. 2.1) of any type may be a 3D monolithic heterostructure that consists of the following regions:

- Concentrator of IR radiation that directs incident radiation onto absorber (an example is an immersion lens made of a wide-gap semiconductor);
- Absorber of IR radiation where the generation of free carriers occurs (this is a narrow-gap semiconductor with bandgap, doping, and geometry selected for the highest ratio of the optical-to-thermal generation rates);
- Contacts to the absorber, which sense the optically generated charge carriers (contacts should not contribute to the dark current of the device; examples are wide-gap heterojunction contacts used in modern devices);
- Passivation of the absorber (the surfaces of the absorber must be insulated from the ambient by a material that also doesn't contribute to carrier generation; in addition, passivation repels the carriers optically generated in the absorber, keeping them away from surfaces where recombination can reduce the quantum efficiency); and
- Retroreflector to enhance absorption (examples are metal or dielectric mirrors; optical resonant cavity structures also can be used).[56, 57]

The structures of an optimized photodetector make it possible to reach the theoretical performance limits in practice. Structures obtained by 3D gap and doping engineering of $Hg_{1-x}Cd_xTe$ are examples of such a device.[58]

Modeling of photodetectors is a strategically important task that is necessary to understand photodetector properties and to optimize their design. Analytical models were developed for specific types of IR devices based on idealized structures, operating both in equilibrium and nonequilibrium mode. These models make some features of the device's operation easy to understand and analyze.

In general, however, the operation of advanced devices can no longer be described by analytical models. Omitting specific features of narrow-gap materials—such as degeneracy and nonparabolic conduction band—may result in enormous errors. The nonequilibrium mode of operation of IR photodetectors brings further complications. The devices are based on an absorber that is near intrinsic or just extrinsic at the operating temperature. The properties of the device differ from those with an extrinsic absorber.[59] First, drift and diffusion are dominated by ambipolar effects due to space charge coupling between electrons and holes. Second, the concentration of charge carriers in near-intrinsic materials can be driven to levels considerably below intrinsic concentrations. As a result, the perturbation can be described only in terms of large signal theory. Third, carrier concentrations in low bandgap materials are dominated by Auger generation and recombination.

An accurate description of more and more complex device architectures, including doping and bandgap grading; heterojunctions; 2D and 3D effects; ambipolar effects; nonequilibrium operation; and surface, interface, and contact effects, can be achieved only with solutions to the fundamental equations that describe the electrical behavior of semiconductor devices. These partial differential equations include continuity equations for electrons and holes and Poisson's equation:

$$\frac{\partial n}{dt} = \frac{1}{q}\vec{\nabla}\cdot\vec{J}_n + G_n - R_n, \tag{4.27}$$

$$\frac{\partial p}{dt} = \frac{1}{q}\vec{\nabla}\cdot\vec{J}_p + G_p - R_p, \tag{4.28}$$

$$\varepsilon_o\varepsilon_r\nabla^2\Psi = -q(N_d^+ - N_a^- + p - n) - \rho_s, \tag{4.29}$$

where Ψ is the electrostatic potential defined as the intrinsic Fermi potential, ρ_s is the surface charge density, and N_d^+ and N_a^- are concentrations of ionized donors and acceptors.[60]

The solution to the equation set (4.27)–(4.29) makes it possible to analyze stationary and transient phenomena in semiconductor devices. The main problem with the solution to these equations is their nonlinearity and complex dependences of their parameters. In many cases, some simplifications are possible. From Boltzmann transport theory, the current densities $\vec{J}_n$ and $\vec{J}_p$ can be written as functions of the carrier concentrations and the quasi-Fermi potentials for electrons and holes, Φ_n and Φ_p:

$$\vec{J}_n = -q\mu_n n\vec{\nabla}\Phi_n, \tag{4.30}$$

and

$$\vec{J}_p = -q\mu_p p\vec{\nabla}\Phi_p. \tag{4.31}$$

Alternatively, $\vec{J}_n$ and $\vec{J}_p$ can be written as functions of Ψ, n, and p, consisting of drift and diffusion components:

$$\vec{J}_n = q\mu_n \vec{E}_e + q D_n \vec{\nabla} n, \qquad (4.32)$$

and

$$\vec{J}_p = q\mu_p \vec{E}_h - q D_p \vec{\nabla} p, \qquad (4.33)$$

where D_n and D_p are the electron and hole diffusion coefficients.

If the effects of bandgap narrowing are neglected and Boltzmann carrier statistics are assumed, then

$$\vec{E}_n = \vec{E}_p = \vec{E} = -\vec{\nabla}\Psi. \qquad (4.34)$$

The steady-state behavior of 1D devices can be described by the set of five differential equations with suitable boundary conditions: two transport equations for electrons and holes, two continuity equations for electrons and holes, and the Poisson equation, which are all related to the Van Roosbroeck:

$$J_n = q D_n \frac{dn}{dx} - q\mu_n n \frac{d\Psi}{dx}, \quad \text{current transport for electrons;} \qquad (4.35)$$

$$J_p = q D_p \frac{dp}{dx} - q\mu_p p \frac{d\Psi}{dx}, \quad \text{current transport for holes;} \qquad (4.36)$$

$$\frac{1}{q}\frac{dJ_n}{dx} + (G - R) = 0, \quad \text{continuity equation for electrons;} \qquad (4.37)$$

$$\frac{1}{q}\frac{dJ_p}{dx} - (G - R) = 0, \quad \text{continuity equation for holes;} \qquad (4.38)$$

$$\frac{d^2\Psi}{dx^2} = -\frac{q}{\varepsilon_o \varepsilon_r}(N_d^+ - N_a^- + p - n), \quad \text{Poisson's equation.}[60] \qquad (4.39)$$

The fundamental equations cannot be solved analytically without the approximations, even for the 1D steady-state case. Therefore, the numerical solutions must be applied. The numerical solution is composed of three steps: (1) grid generation, (2) discretization to transform the differential equations into the linear algebraic equations, and (3) the solution. The Newton direct method is usually used to solve the matrix equation.[62] Other methods are also used to improve the convergence speed and reduce the number of iterations.[63]

Since experiments with complex device structures are complicated, costly, and time consuming, numerical simulations have become a critical tool to develop advanced detectors.[64] Some laboratories have developed suitable software; for example, Stanford University, USA; Military Technical University, Poland;[65] Honyang University, Korea;[63] and others. Commercial simulators are available from several sources, including: Medici (Technology Modeling Associates), Semi-cad (Dawn Technologies), Atlas/Blaze/Luminouse (Silvaco International, Inc.), APSYS (Crosslight Software, Inc.), and others. APSYS, for example, is a full 2D/3D simulator that solves not only the Poisson's equation and the current continuity equations (including such features as field dependent mobilities and avalanche multiplication), but also the scalar wave equation for photonic waveguiding devices (such as waveguide photodetectors) and heat transfer equations with flexible thermal boundary conditions and arbitrary temperature-dependent parameters.

Although existing simulators still do not fully account for all semiconductor properties important for high-temperature photodetectors, they are already invaluable tools for analysis and development of improved IR photodetectors. In addition to device simulators, process simulators are being developed that facilitate the development of advanced device growth technologies.[66, 67]

References

1. R. Dornhaus, G. Nimtz, and B. Schlicht, *Narrow-Gap Semiconductors*, Springer Verlag, Berlin (1983).
2. E. O. Kane, "Band structure of InSb," *J. Phys. Chem. Solids* **1**, 249–261 (1957).
3. W. W. Anderson, "Absorption constant of $Pb_{1-x}Sn_xTe$ and $Hg_{1-x}Cd_xTe$ alloys," *Infrared Phys.* **20**, 363–372 (1980).
4. R. Beattie and A. M. White, "An analytic approximation with a wide range of applicability for band-to-band radiative transition rates in direct, narrow-gap semiconductors," *Semicond. Sci. Technol.* **12**, 359–368 (1997).
5. F. Urbach, "The long wavelength edge of photographic sensitivity and of the electronics of solids," *Phys. Rev.* **92**, 1324 (1953).
6. J. I. Pankove, *Optical Processes in Semiconductors*, Dover, New York (1971).
7. E. Finkman and S. E. Schacham, "The exponential optical absorption band tail of $Hg_{1-x}Cd_xTe$," *J. Appl. Phys.* **56**, 2896–2900 (1984).
8. S. E. Schacham and E. Finkman, "Recombination mechanisms in p-type HgCdTe: Freezout and background flux effects," *J. Appl. Phys.* **57**, 2001–2009 (1985).
9. C. A. Hougen, "Model for infrared absorption and transmission of liquid-phase epitaxy HgCdTe," *J. Appl. Phys.* **66**, 3763–3766 (1989).
10. J. Chu, B. Li, K. Liu, and D. Tang "Empirical rule of intrinsic absorption spectroscopy in $Hg_{1-x}Cd_xTe$," *J. Appl. Phys.* **75**, 1234–1235 (1994).
11. J. Chu, Z. Mi, and D. Tang, "Band-to-band absorption in narrow-gap $Hg_{1-x}Cd_xTe$ semiconductors," *J. Appl. Phys.* **71**, 3955–3961 (1992).
12. R. K. Sharma, D. Verma, and B. B. Sharma, "Observation of below band gap photoconductivity in mercury cadmium telluride," *Infrared Phys. Technol.* **35**, 673–680 (1994).

13. B. Li, J. H. Chu, Y. Chang, Y. S. Gui, and D. Y. Tang, "Optical absorption above the energy band gap in $Hg_{1-x}Cd_xTe$," *Infrared Phys. Technol.* **37**, 525–531 (1996).

14. S. Krishnamurthy, A. B. Chen, and A. Sher, "Near band edge absorption spectra of narrow-gap III-V semiconductor alloys," *J. Appl. Phys.* **80**, 4045–4048 (1996).

15. S. Krishnamurthy, A. B. Chen, and A. Sher, "Electronic structure, absorption coefficient, and Auger rate in HgCdTe and thallium-based alloys," *J. Electron. Mater.* **26**, 571–577 (1997).

16. K. Moazzami, D. Liao, J. Phillips, D. L. Lee, M. Carmody, M. Zandian, and D. Edwall, "Optical absorption properties of HgCdTe epilayers with uniform composition," *J. Electron. Mater.* **32**, 646–650 (2003).

17. K. Moazzami, J. Phillips, D. Lee, D. Edwall, M. Carmody, E. Piquette, M. Zandian, and J. Arias, "Optical-absorption model for molecular-beam epitaxy HgCdTe and application to infrared detector photoresponse," *J. Electron. Mater.* **33**, 701–708 (2004).

18. J. Piotrowski, W. Galus, and M. Grudzień, "Near room-temperature IR photodetectors," *Infrared Phys.* **31**, 1–48 (1990).

19. J. A. Mroczkowski and D. A. Nelson, "Optical absorption below the absorption edge in $Hg_{1-x}Cd_xTe$," *J. Appl. Phys.* **54**, 2041–2051 (1983).

20. C. T. Elliott and N. T. Gordon, "Infrared Detectors," in *Handbook on Semiconductors*, Vol. 4, C. Hilsum (Ed.), North-Holland, Amsterdam, 841–936 (1993).

21. A. Rogalski (Ed.), *Infrared Photon Detectors*, SPIE Press, Bellingham, WA (1995).

22. A. Rogalski, K. Adamiec, and J. Rutkowski, *Narrow-Gap Semiconductor Photodiodes*, SPIE Press, Bellingham, WA (2000).

23. V. C. Lopes, A. J. Syllaios, and M. C. Chen, "Minority carrier lifetime in mercury cadmium telluride," *Semicond. Sci. Technol.* **8**, 824–841 (1993).

24. Y. Yamamoto, Y. Miyamoto, and K. Tanikawa, "Minority carrier lifetime in the region close to the interface between the anodic oxide and CdHgTe," *J. Crystal Growth* **72**, 270–274 (1985).

25. S. M. Johnson, D. R. Rhiger, J. P. Rosbeck, J. M. Peterson, S. M. Taylor, and M. E. Boyd, "Effect of dislocations on the electrical and optical properties of long-wavelength infrared HgCdTe photovoltaic detectors," *J. Vac. Sci. Technol. B* **10**, 1499–1506 (1992).

26. K. Jóźwikowski and A. Rogalski, "Effect of dislocations on performance of LWIR HgCdTe photodiodes," *J. Electron. Mater.* **29**, 736–741 (2000).

27. M. C. Chen, L. Colombo, J. A. Dodge, and J. H. Tregilgas, "The minority carrier lifetime in doped and undoped p-type $Hg_{0.78}Cd_{0.22}Te$ liquid phase epitaxy films," *J. Electron. Mater.* **24**, 539–544 (1995).

28. T. Tung, L .V. DeArmond, R. F. Herald, P. E. Herning, M. H. Kalisher, D. A. Olson, R. F. Risser, A. P. Stevens, and S. J. Tighe, "State of the art of Hg-melt LPE HgCdTe at Santa Barbara Research Center," *Proc. SPIE* **1735**, 109–131 (1992).

29. P. Mitra, T. R. Schimert, F. C. Case, R. Starr, M. H. Weiler, M. Kestigian, and M. B. Reine, "Metalorganic chemical vapor deposition of HgCdTe for photodiode applications," *J. Electron. Mater.* **24**, 661–668 (1995).

30. P. Mitra, Y. L. Tyan, F. C. Case, R. Starr, and M. B. Reine, "Improved arsenic doping in metalorganic chemical vapor deposition of HgCdTe and in situ growth of high performance long wavelength infrared photodiodes," *J. Electron Mater.* **25**, 1328–1335 (1996).

31. M. J. Bevan, M. C. Chen, and H. D. Shih, "High-quality p-type $Hg_{1-x}Cd_xTe$ prepared by metalorganic chemical vapor deposition," *Appl. Phys. Lett.* **67**, 3450–3452 (1996).

32. G. Destefanis and J. P. Chamonal, "Large improvement in HgCdTe photovoltaic detector performances at LETI," *J. Electron. Mater.* **22**, 1027–1032 (1993).

33. C. L. Littler, E. Maldonado, X. N. Song, Z. You, J. L. Elkind, D. G. Seiler, and J. R. Lowney, "Investigation of mercury interstitials in $Hg_{1-x}Cd_xTe$ alloys using resonant impact-ionization spectroscopy," *J. Vac. Sci. Technol. B* **9**, 1466–1470 (1991).

34. M. A. Kinch, "Fundamental physics of infrared detector materials," *J. Electron. Mater.* **29**, 809–817 (2000).

35. M. A. Kinch, F. Aqariden, D. Chandra, P. K. Liao, H. F. Schaake, and H. D. Shih, "Minority carrier lifetime in p-HgCdTe," *J. Electron. Mater.* **34**, 880–884 (2005).

36. R. G. Humpreys, "Radiative lifetime in semiconductors for infrared detectors," *Infrared Phys.* **23**, 171–175 (1983).

37. R. G. Humpreys, "Radiative lifetime in semiconductors for infrared detectors," *Infrared Phys.* **26**, 337–342 (1986).

38. T. Elliott, N. T. Gordon, and A. M. White, "Towards background-limited, room-temperature, infrared photon detectors in the 3–13 μm wavelength range," *Appl. Phys. Lett.* **74**, 2881–2883 (1999).

39. N. T. Gordon, C. D. Maxey, C. L. Jones, R. Catchpole, and L. Hipwood, "Suppression of radiatively generated currents in infrared detectors," *J. Appl. Phys.* **91**, 565–568 (2002).

40. C. T. Elliott and C. L. Jones, "Non-equilibrium Devices in HgCdTe," in *Narrow-gap II-VI Compounds for Optoelectronic and Electromagnetic Applications*, P. Capper (Ed.), Chapman & Hall, London, 474–485 (1997).

41. T. Ashley, N. T. Gordon, G. R. Nash, C. L. Jones, C. D. Maxey, and R. A. Catchpole, "Long-wavelength HgCdTe negative luminescent devices," *Appl. Phys. Lett.* **79**, 1136–1138 (2001).

42. A. R. Beattie and P. T. Landsberg, "Auger effect in semiconductors," *Proc. Roy. Soc. A* **249**, 16–29 (1959).

43. J. S. Blakemore, *Semiconductor Statistics*, Pergamon Press, Oxford (1962).

44. T. N. Casselman and P. E. Petersen, "A comparison of the dominant Auger transitions in p-type (Hg,Cd)Te," *Solid State Commun.* **33**, 615–619 (1980).

45. T. N. Casselman, "Calculation of the Auger lifetime in p-type $Hg_{1-x}Cd_xTe$," *J. Appl. Phys.* **52**, 848–854 (1981).
46. P. E. Petersen, "Auger recombination in mercury cadmium telluride," in *Semiconductors and Semimetals*, Vol. 18, R. K. Willardson and A. C. Beer (Eds.), Academic Press, New York, 121–155 (1981).
47. A. M. White, "The characteristics o minority-carrier exclusion in narrow direct gap semiconductors," *Infrared Physics* **25**, 729–741 (1985).
48. A. Rogalski and J. Piotrowski, "Intrinsic infrared detectors," *Prog. Quant. Electr.* **12**, 87–289 (1988).
49. C. M. Ciesla, B. N. Murdin, T. J. Phillips, A. M. White, A. R. Beattie, C. J. G. M. Langerak, C. T. Elliott, C. R. Pidgeon, and S. Sivananthan, "Auger recombination dynamics of $Hg_{0.795}Cd_{0.205}Te$ in the high excitation regime," *Appl. Phys. Lett.* **71**, 491–493 (1997).
50. S. Krishnamurthy and T. N. Casselman, "A detailed calculation of the Auger lifetime in p-type HgCdTe," *J. Electron. Mater.* **29**, 828–831 (2000).
51. S. Krishnamurthy, M. A. Berding, Z. G. Yu, C. H. Swartz, T. H. Myers, D. D. Edwall, and R. DeWames, "Model of minority carrier lifetimes in doped HgCdTe," *J. Electron. Mater.* **34**, 873–879 (2005).
52. R. Beattie and A. M. White, "An analytic approximation with a wide range of applicability for electron initiated Auger transitions in narrow-gap semiconductors," *J. Appl. Phys.* **79**, 802–813 (1996).
53. T. Ashley and C. T. Elliott, "Non-equilibrium mode of operation for infrared detection," *Electron. Lett.* **21**, 451–452 (1985).
54. T. Ashley, T. C. Elliott and A. M. White, "Non-equilibrium devices for infrared detection," *Proc. SPIE* **572**, 123–132 (1985).
55. S. M. Sze, *Physics of Semiconductor Devices*, Wiley, New York (1982).
56. J. Piotrowski. "$Hg_{1-x}Cd_xTe$: material for the present and future generation of infrared sensors," *MST News POLAND* No. **1**, 4–5, March 1997.
57. J. Piotrowski and A. Rogalski, "New generation of infrared photodetectors," *Sensors and Actuators A* **67**, 146–152 (1997).
58. J. Piotrowski and M. Razeghi, "Improved performance of IR photodetectors with 3D gap engineering," *Proc. SPIE* **2397**, 180–192 (1995).
59. M. White, "Auger suppression and negative resistance in low gap diode structures," *Infrared Phys.* **26**, 317–324 (1986).
60. *MEDICI Manual*, Technology Modeling Associates, Inc., Palo Alto, CA (1994).
61. W. Van Roosbroeck, "Theory of the electrons and holes in germanium and other semiconductors," *Bell Syst. Tech. J.* **29**, 560–607 (1950).
62. M. Kurata, *Numerical Analysis of Semiconductor Devices*, Lexington Books, DC Heath and Co., Lexington, MA (1982).
63. S. D. Yoo, N. H. Jo, B. G. Ko, J. Chang, J. G. Park, and K. D. Kwack, "Numerical simulations for HgCdTe related detectors," *Opto-Electron. Rev.* **7**, 347–356 (1999).

64. K. Kosai, "Status and application of HgCdTe device modeling," *J. Electron. Mater.* **24**, 635–640 (1995).
65. K. Jóźwikowski, "Numerical modeling of fluctuation phenomena in semiconductor devices," *J. Appl. Phys.* **90**, 1318–1327 (2001).
66. J. L. Meléndez and C. R. Helms, "Process modeling and simulation of $Hg_{1-x}Cd_xTe$. Part I: Status of Stanford University mercury cadmium telluride process simulator," *J. Electron. Mater.* **24**, 565–571 (1995).
67. J. L. Meléndez and C. R. Helms, "Process modeling and simulation for $Hg_{1-x}Cd_xTe$. Part II: Self-diffusion, interdiffusion, and fundamental mechanisms of point-defect interactions in $Hg_{1-x}Cd_xTe$," *J. Electron. Mater.* **24**, 573–579 (1995).

Chapter 5

$Hg_{1-x}Cd_xTe$ Photoconductors

The first results on photoconductivity in $Hg_{1-x}Cd_xTe$ were reported by Lawson et al. in 1959.[1] Ten years later, in 1969, Bartlett et al. reported background limited performance of photoconductors operated at 77 K in the 8–14-μm LWIR spectral region.[2] The advances in material preparation and detector technology have led to devices approaching the theoretical limits of responsivity over wide ranges of temperature and background (see Refs. 3–6). The simple n-type $Hg_{1-x}Cd_xTe$ photoconductor is a mature product that has been in production for over 10 years. The largest market was for 60-, 120-, and 180-element units in the "common module" military thermal imaging viewers. Photoconductivity was the most common mode of operation of 3–5-μm and 8–14-μm $Hg_{1-x}Cd_xTe$ photodetectors for many years.

In 1974, Elliott reported a major advance in IR detectors in which the detection, time delay, and integration functions in serial scan thermal imaging systems were performed within a simple three-lead filament photoconductor, known by the acronym SPRITE (Signal PRocessing In The Element).[7]

The further development of photoconductors is connected with the elimination of the deleterious effect of sweep-out[8, 9] by the application of accumulated[10–12] or heterojunction[13, 14] contacts. Heterojuction passivation has been used to improve stability.[15, 16] The operation of 8–14-μm photodetectors has been extended to ambient temperatures.[17–20] The means applied to improve the performance of photoconductors operated without cryogenic cooling include the optimized p-type doping, the use of optical immersion and optical resonant cavities. Elliott and other British scientists introduced Auger-suppressed excluded photoconductors.[21, 22]

The research activity on photoconductors has been significantly reduced in the last decade on reflecting maturity of the devices. At the same time, $Hg_{1-x}Cd_xTe$ photodetectors are still manufactured in large quantities and used in many important applications.

5.1 Simplified Model of Equilibrium Mode Photoconductors

The physics and principle of operation of IR photoconductors are summarized in review papers.[7, 23, 24] This section will discuss primarily the specific features of the photoconductors operating at near room temperatures.

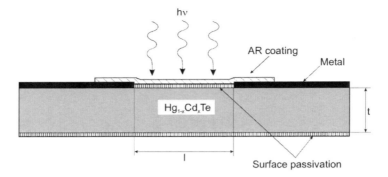

Figure 5.1 Geometry of a photoconductor.

A typical photoconductor is a flake of the narrow gap $Hg_{1-x}Cd_xTe$, rectangular in shape, and supplied with two electric contacts (see Fig. 5.1). Typically, the front and backside surfaces are covered with passivation coating. An antireflection (AR) coating is also frequently used. The detector resistance is usually low (10–300 Ω). At present more complex devices structures are frequently used with heterojunction passivation and heterojunction contacts.[15]

The full analysis of the device can be performed by solving continuity equations with suitable boundary conditions. Analytical solutions exist for homogeneous semiconductors,[23] assuming ohmic contacts and near-equilibrium operation. Even in this case, the expressions are quite complicated.

Here we deliver a simplified approach that describes the properties of photoconductors operating at near room temperatures in which the photoconductor can be described in terms of quantum efficiency, current gain, generation-recombination noise, Johnson-Nyquist noise, and low frequency noise. Since electron mobility is much larger than the hole mobility, the photoconductivity is mainly due to electrons. The photoconductivity gain is the ratio of effective carrier lifetime, τ_{ef}, to the electron transit time, τ_t, that is time of electron drift between electrodes:

$$g = \frac{\tau_{ef}}{\tau_t} = \frac{\tau_{ef}}{l/\mu_e E} = \frac{V \tau_{ef} \mu_e}{l^2}. \qquad (5.1)$$

The current responsivity is

$$R_i = \frac{\lambda}{hc} \eta q g = \frac{\lambda \eta}{hc} \frac{q V \tau_{ef} \mu_e}{l^2}. \qquad (5.2)$$

The voltage responsivity is

$$R_v = R_i R = \frac{\lambda \eta}{hc} \frac{q V \tau_{ef} \mu_e}{l^2} R, \qquad (5.3)$$

where R is the device resistance.

When effective lifetime is larger than the electron drift time, the current gain is larger than unity. The maximum effective lifetime is the bulk minority carrier lifetime. The effective lifetime can be reduced through the recombination of minority carriers at surfaces, interfaces, and contacts. Drift of minority carriers out of the active region to the contact where they recombine is another reason for reduction of effective lifetime (sweep-out effect).[8, 9] The sweep-out effect becomes significant when the ambipolar drift time becomes comparable or shorter than the bulk lifetime. This readily occurs for high bias voltages, short devices, and long bulk lifetimes. The photoconductivity gain versus the applied voltage saturates at the level of the ratio of ambipolar transit time averaged within the active area to the electron transit time. The maximum gain $g_{max} = \mu_e/2\mu_p$ (≈ 50) can be achieved in n-type $Hg_{1-x}Cd_xTe$.

The sweep out effect is more pronounced for p-type material due to a high ambipolar mobility where maximum gains are smaller than $\mu_e/2\mu_p$. The positive feature of sweep-out is an improvement of the high-frequency characteristics due to decreasing effective lifetime.

The total noise voltage of a photoconductor is

$$V_n^2 = V_{GR}^2 + V_J^2 + V_{1/f}^2. \tag{5.4}$$

The generation-recombination noise voltage for equilibrium conditions is

$$V_{GR}^2 = 2(G + R)lwt(Rqg)^2\Delta f, \tag{5.5}$$

where G and R are the volume generation and recombination rates, and lwt is the volume of detector.

The Johnson-Nyquist thermal noise voltage is

$$V_J^2 = 4kTR\Delta f. \tag{5.6}$$

The low frequency noise voltage can be described by the Hooge's expression

$$V_{1/f}^2 = \alpha_H \frac{V^2}{Nf}\Delta f, \tag{5.7}$$

where α_H is the Hooge's constant and N is the number of charge carriers. Frequently, the low frequency is characterized by the $1/f$ noise knee frequency $f_{1/f}$

$$V_{1/f}^2 = V_{GR}^2 \frac{f_{1/f}}{f}. \tag{5.8}$$

The value of the Hooge constant and $f_{1/f}$ is usually considered the technology-related property of the device. There are, however, quantum $1/f$ noise theories describing the $1/f$ noise as the fundamental material property.[24] Hooge's constant in the range 5×10^{-3} to 3.4×10^{-5} have been measured frequently below the lower limit calculated according to existing theories.[25]

At high frequencies the low frequency noise can be neglected and detectivity is determined by the generation-recombination noise and Johnson noise. The $\alpha\tau/n_i$ has been found as material figure of merit for photoconductors, which actually is the α/G figure of merit.[26]

5.1.1 High-frequency operation of photoconductors

The photoconductivity gain remains constant for frequencies smaller than cut-off frequency $f_{co} = 1/2\pi\tau_{ef}$. At higher frequencies, photoconductivity gain decreases:

$$g(f) = \frac{V\tau_{ef}\mu_e}{l^2}\left[1 + \left(\frac{f}{f_{co}}\right)\right]^{-1/2}. \tag{5.9}$$

As a result, the current and voltage responsivities, and generation-recombination noise voltages decrease with frequency in the similar way (see Fig. 5.2). The noise versus the frequency curve shows three characteristic regions: (1) $1/f$ noise region at frequencies below $1/f$ knee with $\sim f^{-1/2}$ slope, (2) white noise generation-recombination (G-R) region, and (3) white noise Johnson-Nyquist region. This results in corresponding regions on D^*-f characteristics. Initially, the detectivity increases with a slope of $f^{1/2}$ achieving the generation-recombination limit at frequencies above the $1/f$ frequency knee, which remains constant until G-R noise dominates over thermal noise and falls down with $1/f$ slope in the Johnson-Nyquist thermal noise region. It is worth noting that the cutoff frequency for detec-

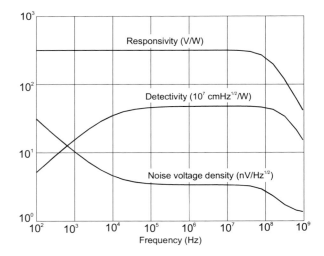

Figure 5.2 Dependences of photoconductor parameters on frequency. Parameters: $x = 0.17$, $N_a = 1 \times 10^{17}$ cm^{-3}, $T = 220$ K, $\lambda = 12$ μm, $V = 0.4$ V, size 50×50 μm, and $f_{1/f} = 10$ kHz.

tivity is larger than the cutoff frequency for photoconductivity gain. This is because the transition from G-R region occurs at frequencies larger than cutoff frequency for the photoconductivity gain.

5.1.2 10.6-µm Hg$_{1-x}$Cd$_x$Te photoconductor operating at 200–300 K

Let us consider the specific example of a 10.6-µm Hg$_{1-x}$Cd$_x$Te photoconductors operated at 200–300 K.[17, 26] In this case, the generation and recombination rates are govern by the Auger processes. This is an example of a device whose performance is well described by the simple model. Typically, diffusion length and ambipolar drift length are shorter than the length of the device (for device lengths of 50 µm or more).

Figure 5.3 shows calculated performance of uncooled 10.6-µm Hg$_{1-x}$Cd$_x$Te photoconductors as a function of doping.[17] In calculations, the bias power density $P/A = 1$ W/mm^2 was assumed not to significantly increase the temperature of the active element. This is attainable only in well-designed heat dissipation devices.

The calculations show clear advantage to fabricate this type of photoconductor from p-type materials. The resistance, recombination time, voltage responsivity, and detectivity increases with a light p-type doping compared to the intrinsic material peaking at various levels of p-type doping and decreasing again with further doping. The increase of resistance is due to reduced concentration of mobile electrons. Resistance achieves its maximum at the highest level of doping $N_a \approx 10 n_i \approx (\mu_e/\mu_h)^{1/2} n_i$ as the result of a high μ_e/μ_h ratio (≈ 100).

The increase of voltage responsivity is due to increased recombination time, absorption coefficient, and resistance, peaking at $z = p/n_i \approx 3$. Further doping re-

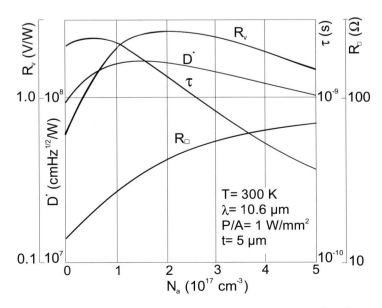

Figure 5.3 Properties of uncooled 10.6-µm photoconductors as a function of doping.

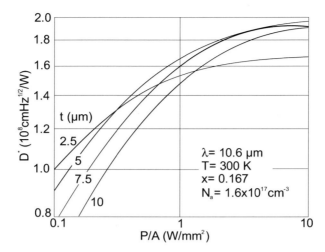

Figure 5.4 Detectivity of uncooled 10.6-μm photoconductors as a function of bias power density.

duces the voltage responsivity mostly because of the reduction of lifetime. Detectivity is peaking at somewhat lower doping compared to the voltage responsivity. This is due to the increase of the A7 generation and recombination rate with the increase of doping above the minimum generation and recombination rate doping level.

Figure 5.4 illustrates the influence of bias power density on the normalized detectivity.[17] The detectivity increases with bias at low fields when the Johnson-Nyquist thermal noise remains significant. For low bias, the optimum thickness is smaller than for large bias. In this case, the loss of photons due to week absorption is compensated by the reduction of noise. At a high bias, the detectivity achieves the generation-recombination noise limit. In this case, the optimum thickness is $1.26/\alpha$ as follows from the general consideration in Chapter 2. Very good bias power dissipation is required to achieve this maximum performance. This may be difficult to realize in practice, however. It should be also noted that the optimum composition, doping, and detector thickness depend upon the bias power density.

Calculations show that the detectivity exceeding 1.6×10^8 cm Hz$^{1/2}$/W can be achieved in optimized devices with 1 W/mm^2 bias power dissipation. In this case, the G-R noise limit of performance is 25% higher. Therefore, uncooled long wavelength photoconductors cannot readily achieve their generation-recombination limits due to heat dissipation constrains. A possible exception is a very small size device ($< 10 \times 10$ μm) where efficient 3D heat dissipation occurs. The strong bias may result in a high $1/f$ noise, however.

As Fig. 5.5 shows, the performance of 10.6-μm photoconductors can be significantly improved by a moderate cooling, which can be conveniently done with one- or several-stage thermoelectric (TE) coolers.[17] Due to increased photoelectric gain, the generation-recombination limit of performance is readily achievable in TE cooled $\lambda \approx 10$ μm photoconductors. At 200 K, the calculated detectivity is

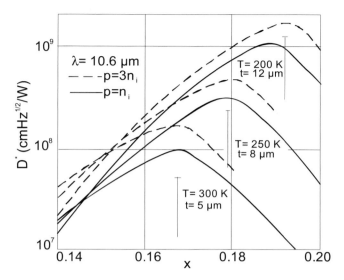

Figure 5.5 Detectivity of 10.6-μm photoconductors as a function of composition, temperature, and doping. The measured values are indicated by the vertical lines.

$\approx 2 \times 10^9$ cm Hz$^{1/2}$/W. However, the improvement of performance is accompanied by an increase in response time as the result of increasing lifetime of charge carriers.

A careful selection of composition is required for the best performance at a specified temperature. For example, the 10.6-μm devices with optimum composition for 200 K operation are practically not sensitive at 10.6 μm when operated at a temperature of 300 K.

It should be noted that the bias requirements for operation of the LWIR photoconductors results in a significant heat load. This prevents practical use of photoconductors that are large area devices (single element or arrays) with or without TE cooling.

Numerous papers have reported detailed analyses of specific examples of low-temperature 8–14-μm and 3–5-μm devices.[3–6, 24, 25, 27–33] It has been well established that the generation-recombination processes in high-quality n-type devices are dominated by the A1 mechanism. Background radiation has a decisive influence on performance since the concentration of both majority and minority carriers in 80 K 8–14-μm devices and the concentration of minority carriers in 3–5-μm devices are typically determined by background flux.

While cooling and/or reduction of the cut-off wavelength effectively increases the fundamental limits of performance, these processes also makes the device more vulnerable to nonfundamental limitations. This is the reason why, in contrast to near room temperatures LWIR photoconductors, the SWIR, MWIR, and LWIR devices operated around 80 K are typically based on n-type $Hg_{1-x}Cd_x$Te. Good reproducibility of low-level ($\approx 10^{14}$ cm^{-3}) doping, which is required for the best performance at low temperatures is possibly the most important advantage of n-type material. The low hole diffusion coefficient makes n-type devices less vulnerable to contact and surface recombination. N-type materials also exhibit a lower concentration of Shockley-Read centers. In addition, with n-type materials, good methods of surface passivation are available.

The length of the photoconductors being used in high-resolution thermal imaging systems (≈ 50 µm) is typically less than the minority carrier diffusion and minority carrier drift length in cooled $Hg_{1-x}Cd_xTe$, which causes a pronounced sweep out effect.[25] Sweep-out has been recognized as a major limitation to the 8–14-µm photoconductor performance when they are short and operated at low temperatures with low background radiation. The influence of sweep-out is even stronger for shorter wavelength devices. For $\lambda_{co} < 5$ µm photoconductors with low G-R rates, these effects become significant at low and elevated temperatures, even for relatively long devices, making Johnson-Nyquist/sweep-out's devices limited.

Various approaches to reduce the undesirable sweep-out effects have been suggested. The n^+-n contact can be used to isolate regions containing minority holes from high recombination rate regions.[10–13] Due to degeneracy in the n^+ region the potential barrier to holes is much larger than the simple Boltzmann factor $(kT/q)\ln(n/n_o)$. The effective recombination velocity at the blocking contact below 100 cm/s can be achieved. For this, the electron concentration must be degenerate within the distance of 1 µm or less.

The carrier sweep-out also can be virtually eliminated with a heterojunction contact photoconductor.[13, 14, 31, 32] The heterojunction contacts allow more heavily doped material to be used to make good photoconductors. Rigorous treatment of the blocking contact or heterojunction contact photoconductors is possible only numerically since the concentrations of both the minority and majority carriers in blocking or heterojunction contact fall below their equilibrium values.[34]

The accumulation at surfaces for reflection of minority carriers is equally applicable to reducing the recombination at the active and backside surface of photoconductors. Excessive accumulation, on the other hand, can lead to a large shunt conductance, which also can degrade detectivity.[35] Heterojunction passivation with lightly doped widegap material seems to be more useful.[15, 30]

5.2 Excluded Photoconductors

The excluded photoconductor was proposed for the first time by Elliott and other English workers, and it was the first demonstrated as a practical non-equilibrium device.[7, 21, 22, 36, 37] The principle of operation excluding the contact photoconductor is shown in Fig. 5.6. The positively biased contact is a highly doped or widegap material while the photosensitive area is a near-intrinsic n-type material. Such a contact does not inject minority carriers, but it permits the majority electrons to flow out of the device. As a result, the hole concentration in the vicinity of the contact is decreased and the electron concentration also fails to maintain electroneutrality in the region. Consequently, the Auger generation and recombination processes become suppressed in the excluded zone. The device must be longer than

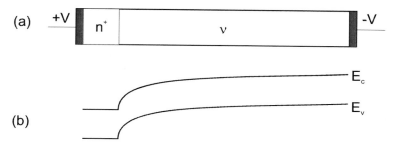

Figure 5.6 (a) Schematic of an excluded contact photoconductor structure; (b) energy levels diagram. (Reprinted from Ref. 7 with permission from Elsevier.)

the exclusion length to avoid the effect of carrier accumulation at the negative bias contact. The excluded zone can be used as the active region of the device.

The accurate analysis of nonequilibrium devices can be done only with the numerical solution of the full continuity equation for electrons and holes.[38–41] The usual approximations break down in the case of non-equilibrium mode of operation due to large departures from the equilibrium.

Figure 5.7 shows an example of the calculated carrier distribution across an n^+-ν excluded device.[40] At high bias current density, the concentrations of electrons and holes in the exclusion region fall from the equilibrium near-intrinsic values to much lower ones. The concentration of electrons achieves its extrinsic value while the hole concentration is even lower. The length of the excluded region depends on the bias current density, bandgap, temperature, and other factors. Lengths higher than 100 μm have been observed experimentally for MWIR

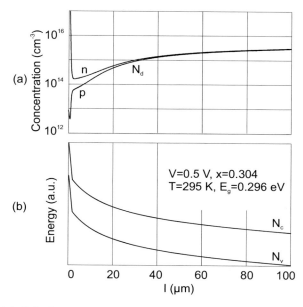

Figure 5.7 (a) Calculated electron and hole concentration, and (b) energy levels.

devices.[37] A threshold current is required to counter the back diffusion current between unexcluded and excluded regions. Thereafter, the resistance rises rapidly as the length of the excluded region increases. As a result, the current-voltage characteristic exhibits saturation for above-threshold currents.

Assuming that the Auger generation is reduced below that of other residual processes, and under some simplifications, analytic expressions have been derived for

- The critical field, where exclusion terminates:

$$E_c = \left(\frac{kT}{q} \frac{G}{r \mu_p N_d} \right)^{1/2} ; \qquad (5.10)$$

- The length of the exclusion zone:

$$L = \frac{\mu_p J}{\mu_n q G} - \left(\frac{D_n \mu_p N_d}{r \mu_n G} \right)^{1/2} ; \quad \text{and} \qquad (5.11)$$

- The threshold bias current J_o:

$$J_o = q \left(\frac{D_n \mu_n G N_d}{r \mu_p} \right)^{1/2} , \qquad (5.12)$$

where r is some small fraction found by numerical calculations over a wide range of conditions of doping, temperature, and bias to be 0.012 ± 0.002, and G is the residual generation rate in the exclusion zone.[22, 42] It should be noted that the length of exclusion zone might exceed the diffusion length.

The exclusion in p^+-π structures can be modeled in a similar manner. But, there are two important differences. First, due to low hole mobility, the majority carriers no longer determine the current flow. Second, heavily doped p^+-contacts exhibit large G-R velocity. This can be solved by using a p-type heterojunction contact with increased bandgap.

The practical realization of nonequilibrium devices depends on several important limitations. The electric field in an excluded region must be sufficiently low to avoid heating a device as a whole and electrons above the lattice temperature. The heating of the structure can be prevented by a proper heat sink design, and it seems not to be a serious limitation—at least for a single element and low-area devices. Electron heating sets a maximum field that has been estimated as 1000 V/cm in materials for uncooled 5-µm devices and a few hundred V/cm in materials for $\approx$ 10-µm devices operated at 180 K. Electron heating is not an important constraint in the 3–5-µm band, but it seems to restrict the usefulness of exclusion at 10.6 µm and longer wavelengths in the 8–14-µm band. Very low doping ($< 10^{14}$ cm^{-3}) is required for effective exclusion; however, the values 3×10^{14} cm^{-3} are typical in industry. Exclusion may be inhibited by any non-Auger generation such as the Shockley-Read or surface generation. A large electrical field may result in flicker noise.

5.3 Practical Photoconductors

5.3.1 Technology of $Hg_{1-x}Cd_xTe$ photoconductors

Photoconductors can be prepared either from $Hg_{1-x}Cd_xTe$ bulk crystals or epilayers. Figure 5.8 shows two typical structures of photoconductors.[43] The main part of all structures is a 3–20-μm flake of $Hg_{1-x}Cd_xTe$ supplied with electrodes. The optimum thickness of the active elements (a few μm) depends upon the wavelength and temperature of operation and is smaller in uncooled long-wavelength devices. The front-side surfaces of both structures are usually covered with a passivation layer and AR coating. The backside surface of the device is also passivated. In contrast, the backside surface of the epitaxial layer grown on the CdZnTe substrate does not require any passivation since the increasing bandgap prevents the reflection of minority carriers. The devices are bonded to heat-conductive substrates.

To increase the absorption of radiation, the detectors are frequently supplied with a gold back reflector[17, 24] and insulated from the photoconductor with a ZnS layer or substrate. The thickness of the semiconductor and the two dielectric layers are frequently selected to establish optical resonant cavities with standing waves in the structure, with peaks at the front and nodes at the back surfaces. For effective interference, the two surfaces must be sufficiently flat.

Numerous fabrication procedures are being used by various manufacturers.[17, 24, 28, 43–45] Technological methods of modern microelectronics are used in $Hg_{1-x}Cd_xTe$ photoconductor manufacturing; however, extreme care is necessary to prevent any mechanical and thermal damage of the material. Fabrication usually starts by selecting the starting materials, which are $Hg_{1-x}Cd_xTe$ wafers or

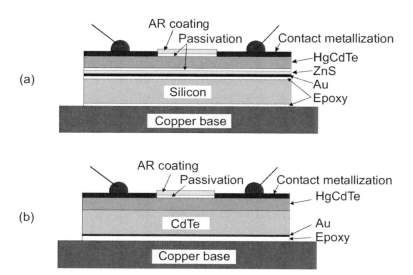

Figure 5.8 Structure of photoconductors based on (a) bulk, and (b) epilayer $Hg_{1-x}Cd_xTe$. The devices are supplied with a back reflector.

epilayers. The typical selection parameters are composition, doping, and minority lifetime mapping.

The key processing steps for manufacturing $Hg_{1-x}Cd_xTe$ photoconductors from bulk $Hg_{1-x}Cd_xTe$ include:

1. Preparation of the backside surface of the $Hg_{1-x}Cd_xTe$ wafer. This procedure involves careful polishing of one side of the $Hg_{1-x}Cd_xTe$ slab with fine (0.3–1 μm) alumina powder, cleaning in organic solvents followed by etching in 1–10% bromine in methanol solution for several minutes, and washing in methanol. Alternatively, various chemical-mechanical polishing procedures can be used. The backside preparation is completed with passivation, which differs for n- and p-type materials.
2. Bonding to a substrate. The most common substrates for bulk-type photoconductors are sapphire, Ge, Irtran 2, Si, and alumina ceramics. Epoxy resin is generally used to bond $Hg_{1-x}Cd_xTe$ wafers to substrates. The thickness of the epoxy layer should be kept below 1 μm for good heat dissipation.
3. Thinning the slab to its final thickness and preparation of the front-side surface. This is done by lapping, polishing, and etching, and then by surface passivation and application of an antireflection coating (usually ZnS). The individual elements are then delineated with wet or dry etching using photolithography. Often side walls of active elements are also passivated.
4. Electric contact preparation. Vacuum evaporation, sputtering, electroplating, and galvanic or chemical metallization after further photolithography is used. External contacts are supplied by gold wire ultrasonic bonding, conductive epoxy bonding or by soldering with indium. Expanded contact pads are sometimes used to prevent damage of the semiconductor.

The preparation of a photoconductor from epilayers is more straightforward because no laborious thinning to a very low thickness and no backside preparation is required. The use of ISOVPE,[17, 46] LPE,[47–50] MOCVD,[18–20, 51–54] and MBE[55] photoconductors have been reported. CdZnTe, which is typically used for substrates, has relatively poor thermal conductivity. Therefore, for the best heat dissipation, the substrates must be thinned below 30 μm and the photoconductors must be fixed to a good heat-sinking support. The heat dissipation is much easier in small-sized devices ($< 50 \times 50$ μm^2) in which 3D dissipation is significant and no thinning of substrates is necessary. Another solution is to use epilayers deposited on good heat-conducting materials such as sapphire, Si, and GaAs.

Low-temperature epitaxial techniques make it possible to grow complex multilayer photoconductor structures that can be used as multicolor devices or devices with a shaped spectral response.[18, 54]

Passivation is one of the most critical steps in the preparation of photoconductors. The passivation must seal the semiconductor, stabilizing it chemically, and often it also acts as an antireflection coating. An excellent review of $Hg_{1-x}Cd_xTe$ native and deposited insulator layers has been published by Nemirovsky and Bahir.[56]

Passivation of n-type materials is commonly performed with the use of anodic oxidation in 0.1 N KOH in a 90% water solution of ethyl glycol.[57–59] Typically, 100-nm-thick oxide layers are grown. Good interface properties of the n-$Hg_{1-x}Cd_xTe$-oxide interface are due to the accumulation (10^{11}–10^{12} electrons per cm^{-2}) of the semiconductor surface during oxidation. Passivation by pure chemical oxidation in an aqueous solution of $K_3Fe(CN)_6$ and of KOH is also used.[60] Dry methods of native oxide growing have been attempted, such as plasma[61] and photochemical oxidation.[62] Passivation can be improved by overcoating with ZnS or SiO_x layers.[63] Another approach to passivation is based on the direct accumulation of the surface to repel minority holes by a shallow ion milling.[64, 65]

Passivation of p-type materials is of strategic importance for near room temperature devices based on p-type absorbers. But passivation still presents practical difficulties. Oxidation is not useful for p-type $Hg_{1-x}Cd_xTe$ because it causes inversion of the surface. In practice, sputtered or electron-beam evaporated ZnS, with the option of a second layer coating,[66] is usually used for passivation of p-type materials. Native sulfides[67] and fluorides[68] also have been proposed.

The use of CdTe for passivation is very promising since it has high resistivity, is lattice matched, and is chemically compatible to $Hg_{1-x}Cd_xTe$.[69–71] Excellent passivation can be obtained with a graded CdTe-$Hg_{1-x}Cd_xTe$ interface.[66] Barriers can be found in both the conduction and valence band. The best heterojunction passivation can be obtained during epitaxial growth.[72] Directly grown *in situ* CdTe layers lead to a low fixed-interface charge. The indirectly grown CdTe passivation layers are not as good as directly grown layers, but are acceptable in some applications. A low thickness (10 nm) of CdTe is recommended in some papers to prevent $Hg_{1-x}Cd_xTe$ lattice stress.[55]

Contact preparation is another critical step. Evaporated In has been used for a long time for contact metallization to n-type material.[24, 45] Multilayer metallization, Cr-Au, Ti-Au, and Mo-Au are used more frequently at present. Metallization is often preceded by a suitable surface treatment. Ion milling was found very useful to accumulate n-type surfaces, and it seems to be the most preferable surface treatment prior to metallization of n-type material. Chemical and dry etch are also used. Preparation of good contacts to p-type material is more difficult. Evaporated, sputtered, or electroless deposited Au and Cr-Au are the most frequently used for contacts to p-type materials.

5.3.1.1 Practical excluded photoconductors

Practical excluded $Hg_{1-x}Cd_xTe$ photoconductors have been reported by English workers.[42] In contrast to the equilibrium mode photoconductors that are usually based on extrinsically p-type doped material, the excluded devices are fabricated from very low-concentration n-type bulk $Hg_{1-x}Cd_xTe$, with the n^+ regions formed by ion milling or degenerate extrinsic doping. The device is schematically shown in Fig. 5.9. The device is longer than the required detector length to avoid the effect of carrier accumulation at the negative contact. The optically sensitive area is defined by means of an opaque mask, and a side-arm potential probe is used for a readout

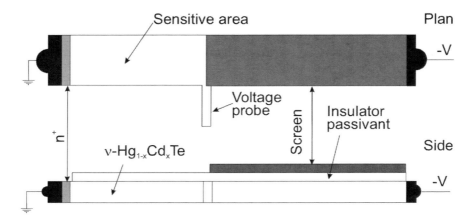

Figure 5.9 Schematic of excluded $Hg_{1-x}Cd_x Te$ photoconductor. (Reprinted from Ref. 37 with permission from IOP.)

contact. This limits the size of the active area to the highly depleted region, which prevents thermal generation in the nondepleted part and at the negative contact from contributing to the noise measured at the readout electrode. ZnS is used for passivation because the usual native oxide passivation produces an accumulated surface, which shunts the excluded region.

5.3.1.2 Optically immersed photoconductors

There are several approaches to fabrication of optically immersed photodetectors. In the conventional hybrid approach, the sensitive elements of photoconductors are cemented with a transparent glue to a prepared lens. For the best result, the immersion lens materials should have a high refraction coefficient and low absorption at the wavelengths of interest. Reflection losses can be prevented by using a suitable antireflection lens coating. Germanium ($n = 4$) is typically used for immersion lenses that can operate both in the 3–5-μm and 8–14-μm bands.[73] Silicon, which is cheap and stable due to its lower refraction coefficient ($n \approx 3.3$), can be used for hyperhemispherical lenses in the 3–5-μm band.[74] The fastest optical system with a Si hyperhemispherical lens is $f/2.2$. The main problem with the conventional approach is developing cement that has suitable optical and mechanical properties. It should be transparent in the required spectral band and it must provide a mechanically stable contact. The refractive coefficient of the cement must be close to that of one of the lenses to prevent refraction losses and to avoid the limitation of a detector acceptance angle. It is not easy to select cement that simultaneously satisfies all these requirements. The use of organic glues has proved to be only partially successful.

$Hg_{1-x}Cd_x Te$ photoconductors coupled with Si lenses through optical contact rather than by mechanical contact have been reported.[74] The detector was glued to the lens, leaving a small (0.2–0.5 μm) gap. Since the gap thickness is significantly less than one wavelength, optical tunneling is possible. The detectors exhibited

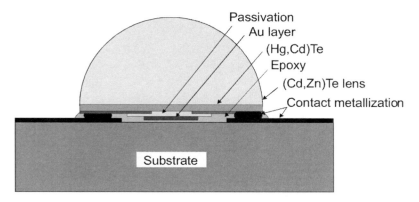

Figure 5.10 Structure of an advanced monolithic optically immersed photoconductor. The active element is supplied with a gold back reflector.

good mechanical stability, and they also achieved a gain that was approximately 20% lower than theoretically possible due to losses in reflection and tunneling. A promising approach is to form lenses from substrates of epilayers.

Monolithic optically immersed photodetectors with lenses formed directly from the transparent CdZnTe ($n = 2.7$) substrate of the $Hg_{1-x}Cd_x$Te layer (Fig. 5.10) have been reported.[17, 75] The relatively low refraction coefficient of CdZnTe makes the expected gain lower than Ge, but hyperhemispherical CdZnTe immersion lenses can be used in relatively fast optics, achieving an approximately 7-fold increase in detectivity and an approximately 50-fold decrease in bias power dissipation. The monolithic approach permits simple and economical manufacturing. These devices are rugged and mechanically stable, and they also can operate in a very broad spectral band with minimal reflection and absorption losses. A similar approach has been used for $Hg_{1-x}Cd_x$Te epilayers deposited on higher refraction coefficient substrates (e.g., GaAs).[18]

The use of the one-lens immersion technology is limited to the fabrication of single-element detectors or arrays of relatively small size because the lens must be larger than the apparent size of the detector to prevent excessive optical distortion of the image. This limitation can be overcome by using arrays of microlenses with one lens for each detector. The integrated detector-microlens array seems to be extremely promising for future high-quality and lightweight thermal imagers based on near room temperature photodetectors. Practical HgCdTe detector arrays with Si microlenses prepared by the use of photolithography have been reported.[76] In contrast to Si, the use of CdZnTe lenses offers a better match to HgCdTe and perfect transparency in a very wide spectral band (1–30 µm). The combination of dry ion milling etch and wet chemical etch has been used at Vigo System S.A. to fabricate 2D arrays of immersion microlenses (see Fig. 5.11).[77]

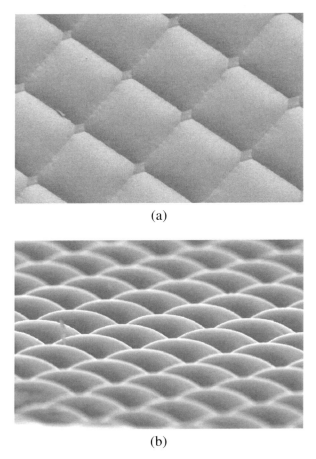

(a)

(b)

Figure 5.11 SEM images of the microlens array: (a) top, and (b) tilted side view. The size of the individual lens is 50×50 μm.

5.3.1.3 Detector housing

The sensitive elements are housed in various packages depending on their operating temperature, designation, and other factors (Fig. 5.12). Ambient temperature detectors, if properly passivated, can operate in an atmospheric ambient, and no window is required. Sealed housings are necessary for cooled devices to prevent water vapor condensation, deposition, and surface contamination. Dry nitrogen, argon, or Kr-Xe mixtures (which have higher thermal conductivity) are frequently used for two- and one-stage thermoelectric coolers. Devices cooled with three-stage coolers are often housed in packages filled with dry Xe, which is both an inert and a low thermal conductivity gas. The < 200 K temperature devices are typically operated in a vacuum. TE-cooled devices are typically supplied with a temperature sensor (thermistor or Si diode) to control the detector temperature.

Special packages are designed to minimize parasitic impedances since the ambient-temperature photoconductors are capable of high-frequency operation

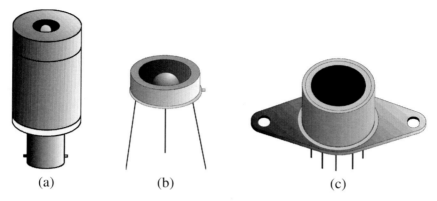

(a)   (b)   (c)

Figure 5.12 Optically immersed MWIR and LWIR $Hg_{1-x}Cd_xTe$ photodetectors: (a) large-area ambient temperature photoconductor; (b) ambient-temperature photoconductor; (c) Peltier-cooled photoconductor and photovoltaic detector. (Reprinted with permission from a Vigo System S.A. data sheet, 2000.)

with typical bandwidths of ≈ 200 MHz, achieving ≈ 1 GHz in high-frequency optimized devices.

5.3.2 Measured performance

One-element, quadrant, and array devices with element sizes ranging from 0.02×0.02 mm^2 to 10×10 mm^2 are being manufactured. Large arrays with the number of elements exceeding 1000 have been reported.[78] Multispectral single elements and arrays also have been developed.[79] The operability of array elements is frequently 100%.

5.3.2.1 Near room temperature devices

The measured detectivities of photoconductors operating at ambient temperature or at temperatures achievable with thermoelectric coolers (200–250 K) are shown in Fig. 5.13. These devices have been fabricated from ISOVPE epilayers, *in situ* doped with foreign impurities. Recently the performance of these devices has been significantly improved due to careful optimization of the compositional and doping profile, the use of metal-back reflectors, better surfaces, and contacts processing.[18–20] The highest measured detectivity of an uncooled and nonimmersed 10.6-μm photoconductor was approximately 7.5×10^7 cm Hz$^{1/2}$/W at high frequency (0.1–100 MHz), which is a factor of ≈ 2 below the theoretical limit.

The present high-temperature photoconductors have relatively poor low-frequency properties. Typically, the $1/f$ knee frequencies of 10 kHz have been observed in uncooled 10.6-μm detectors at electrical fields of ≈ 40 V/cm (typical large-sized Vigo R005 photoconductors). The Hooge's constant of approximately $\approx 10^{-4}$ has been deduced from the low field measurements. The rapid, nonlinear increase of Hooge's constant has been observed for stronger electric fields. Therefore, $1/f$ noise is an issue in near room temperature photoconductors. This

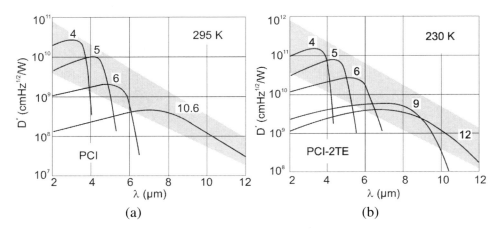

Figure 5.13 Spectral detectivities of optically immersed $Hg_{1-x}Cd_x$Te photoconductors: (a) uncooled, and (b) cooled with two-stage Peltier coolers. Recently, the operation of two-stage cooled photodetectors manufactured at Vigo System S.A. has been extended to ≈ 16 μm with detectivities of $\approx 2 \times 10^9$ cm Hz$^{1/2}$/W at 12 μm.[78] (Reprinted with permission from a Vigo System S.A. data sheet, 2000.)

is especially problematic for uncooled devices, which would require high biasing (≈ 100 V/cm) to approach the generation-recombination performance limit. Cooling to ≈ 200 K reduces the Hooge's constant by a factor of ≈ 2, which in conjunction with much lower electrical field requirements makes it possible to achieve the G-R limited operation at frequencies of ≈ 1000 Hz and higher. Proportionality between G-R and $1/f$ noise in $x = 0.2$ $Hg_{1-x}Cd_x$Te photoconductors has been observed in a wide temperature range (77–250 K).[80]

The reason for poor low-frequency performance is not fully understood at present; inadequate surface passivation and contact technology is usually blamed. No single theory can explain qualitatively the observed low frequency noise.[25]

Optically immersed uncooled and TE-cooled 3–5-μm and 8–12-μm photoconductors have been reported.[17–20, 75] These devices and their spectral detectivities are shown in Fig. 5.13. Significant improvement can be seen in comparison to the non-immersed photodetectors. The devices are especially suitable for high-frequency operation due to a short lifetime, an absorber resistance close to 50 ohms, negligible series resistance, and very low capacitance. Measurements using free-electron lasers revealed response times of ≈ 0.6 ns and ≈ 4 ns at 300 K and 230 K, respectively, for detectivity-optimized 10.6-μm devices. A shorter response of ≈ 0.3 ns was observed in specially designed devices of small physical size ($\approx 10 \times 10$ μm^2 or less) and heavier p-type doping.

The gain-bandwidth product exceeding 1 GHz has been measured in two-stage Peltier cooled 100×100 μm apparent optical area device, optically immersed to GaAs hyperhemispherical lens.

It should be noted that the measured detectivity of uncooled ≈ 10-μm photoconductors is many orders of magnitude higher than other ambient temperature

10.6-µm detectors with subnanosecond response time, such as photon drag detectors, fast thermocouples, bolometers, and pyroelectric ones.

Uncooled two-color photoconductors with individual access to each color region have been reported.[54] The devices were grown with MOCVD. The $Hg_{1-x}Cd_x Te$ layers are either detectors or IR filters while CdTe layers serve as insulators separating various spectral regions.

An interesting device is a two-lead devices that operate over a wide spectral band with performance improved by a large factor at a given short wavelength ranges.[18] The device consists of several stacked active regions (absorbers) with their outputs connected in parallel so the resulting output signal current is the sum of the signals generated at all active regions. Due to a high photoelectric gain in the wider gap absorbers and low thermal generation and recombination, the devices offer significantly better performance at short wavelengths while the long wavelength response remains essentially unaffected. An example is an uncooled photoconductor operating up to 11 µm, with response at 0.9–4 µm increased by $\approx$ 3 orders of magnitude in comparison to the response of conventional 11 µm device.

5.3.2.2 Low-temperature devices

Specific examples of 0.1 eV lightly doped n-type $Hg_{1-x}Cd_x Te$ at 77 K have been reported in numerous papers.[3–6] Devices approaching BLIP performance limits have been fabricated of bulk $n = (2 - 5) \times 10^{14}$ cm^{-3} $Hg_{0.795}Cd_{0.205}Te$, passivated with native oxide and coated with a ZnS AR layer.

Figure 5.14 shows the calculated and measured low-background responsivity and detectivity of a photoconductor as a function of temperature. The generation and recombination rates are clearly dominated by the A1 mechanism. The 77 K de-

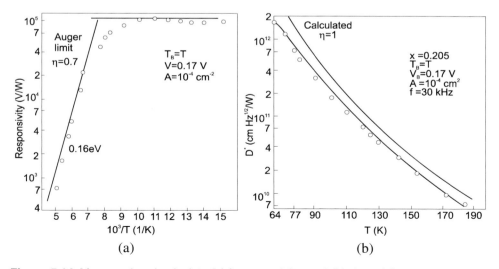

Figure 5.14 Measured and calculated (a) responsivity, and (b) detectivity vs. temperature for a $Hg_{0.795}Cd_{0.205}Te$ photoconductor. (Reprinted from Ref. 4 with permission from Elsevier.)

tectivity achieves a value of approximately 10^{12} cm Hz$^{1/2}$/W, closely approaching the limits predicted by theory.

The nonequilibrium carrier concentration influencing the recombination time and related detector parameters are generated by 300 K background photon flux.[6] The density of background-generated holes, and also electrons for high fluxes, may dominate the thermally generated carriers and decrease the recombination time. This is detrimental to responsivity. The effects of background radiation tend to override any nonuniformities that might be present in the bulk material with regard to element resistance, responsivity, and noise. Vacuum bakeout stability has been checked up to 125°C with little or no change in the vital photoconductive parameters.

At present commercially available 3–5-μm 80 K photoconductors also exhibit background-limited performance.[78] Near-BLIP performance also can be achieved at elevated temperatures up to approximately 200 K.

Little attention has been given to optical immersion of 77 K photoconductors. Such detectors are usually BLIP-limited. Immersion offers no advantages with the exception of some specific cases of large-area devices that otherwise are usually non-BLIP, long-wavelength (with $\lambda_c > 14$ μm), fast-response devices. The use of immersion with such devices facilitates a sweep-out operation mode to extend the frequency response. Large-area (up to 1 cm^2) Hg$_{1-x}$Cd$_x$Te photoconductors with Ge immersion lenses have been reported.[73]

5.3.2.3 Excluded photoconductors

The performance of excluded MWIR photoconductors has been reported by a British team.[22, 37] As Fig. 5.15 shows, the reverse bias responsivity in such a device exhibits a rapid rise as the bias current increases and achieves maximum, which is a factor of tens higher than those at the same direct bias current. The rise is due to increased impedance in the excluded region and to an increase in the effective carrier lifetime to the transit time.

The improvement of detectivity is more modest due to high flicker noise levels at reverse bias. Reversing the bias direction from direct to reverse has improved detectivity by a factor of ≈ 3. This can be related to the poor Hg$_{1-x}$Cd$_x$Te-ZnS interface properties and to fluctuations in the injection rate. An uncooled 10-μm by 10-μm photoconductor with 4.2-μm cutoff has exhibited a 500 K blackbody voltage responsivity of 10^6 V/W, detectivity of 1.5×10^9 cm Hz$^{1/2}$/W at a modulation frequency of 20 kHz, and the thermal figure of merit M^* (zero range, 295 K) of 1.5×10^5 cm^{-1} Hz$^{1/2}$ K^{-1}, which exceeds the performance obtained from any other IR detector operated with the same conditions.

No practical 8–14-μm excluded photoconductors have been demonstrated to date. This can be attributed to the heating of electrons at the high electric fields required for exclusion.

Large detectivity gains could be expected in the future with improvement in the device technology. For example, the performance of the excluded photoconductors

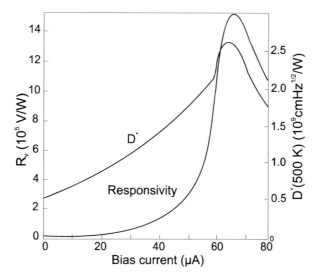

Figure 5.15 D^* (500 K, 20 kHz) and responsivity (500 K) for a 10-μm square Hg$_{1-x}$Cd$_x$Te photoconductor vs. bias current, where $T = 295$ K, and $\lambda = 4.2$ μm. (Reprinted from Ref. 37 with permission from IOP.)

can be improved further by the use of optical immersion and optical resonant cavity. In addition to improved performance due to immersion itself, the immersion increases the apparent size of the small near-contact regions where the exclusion is very effective.

5.4 SPRITE Detectors

The SPRITE detector was originally invented by T. C. Elliott and developed further almost exclusively by British workers.[81–86] This device has been employed in many imaging systems. Figure 5.16 shows the operating principle of the device. The device is essentially an n-type photoconductor with two bias contacts and a readout potential probe. The typical sizes are ≈ 1 mm length, 62.5-μm width, and 10-μm thickness. The device is constant-current biased with the bias field, E, set such that the ambipolar drift velocity v_a, which approximates to the minority hole drift velocity v_d, is equal to the image scan velocity v_s along the device. The length of the device, L, is typically close to or larger than the drift length, $v_d\tau$, where τ is the recombination time.

Consider now an element of the image scanned along the device. The excess carrier concentration in the material increases during the scan, as illustrated in Fig. 5.16. When the illuminated region enters the readout zone, the increased conductivity modulates the output contacts and provides an output signal.

The integration time approximates the recombination time, τ, for long devices. It becomes much longer than the dwell time τ_{pixel} on a conventional element in a fast-scanned serial system. Thus, a proportionally larger (τ/τ_{pixel}) output signal is

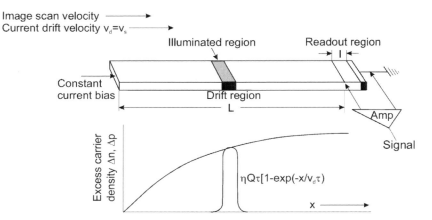

Figure 5.16 Principle of operation of the SPRITE detector. (Reprinted from Ref. 81 with permission from Elsevier.)

observed. If Johnson noise or amplifier noise dominates, it leads to a proportional increase in the signal-to-noise ratio (SNR) with respect to a discrete element. In the background-limited detector, the excess carrier concentration due to background flux also increases by the same factor, but corresponding noise is proportional only to the integrated flux. As a result, the net gain in the SNR with respect to a discrete element is increased by a factor of $(\tau/\tau_{pixel})^{1/2}$.

The voltage responsivity is

$$R_v = \frac{\lambda}{hc} \frac{\eta \tau E l}{n w^2 t}\left[1 - \exp\left(-\frac{L}{\mu_a E \tau}\right)\right], \tag{5.13}$$

where l is the readout zone length, and L is the drift zone length.[81]

The dominant noise is the generation-recombination noise due to fluctuations in the density of thermal and background-radiation generated carriers. The spectral density of noise at low frequencies is

$$V_n^2 = \frac{4E^2 l \tau}{n^2 w t}\left(p_o + \frac{\eta Q_B \tau}{t}\right)\left(1 - \exp\frac{-L}{\mu_a E \tau}\right)\left[1 - \frac{\tau}{\tau_a}\left(1 - \exp\frac{-\tau_a}{\tau}\right)\right]. \tag{5.14}$$

For a long background-limited device in which $L \gg \mu_a E \tau$ and $\eta Q_B \tau/t \gg p_o$,

$$D^* = \frac{\lambda \eta^{1/2}}{2hc}\left(\frac{l}{Q_B w}\right)^{1/2}\left[1 - \frac{\tau}{\tau_a}\left(1 - \exp\frac{-\tau_a}{\tau}\right)\right]^{-1/2}, \tag{5.15}$$

and at sufficiently high speeds such that $\tau_a \ll \tau$, and

$$D^* = (2\eta)^{1/2} D^*_{\text{BLIP}}\left(\frac{l}{w}\right)^{1/2}\left(\frac{\tau}{\tau_a}\right)^{1/2}. \tag{5.16}$$

For a nominal resolution size $w \times w$, the pixel rate is v_a/w and

$$D^* = (2\eta)^{1/2}D^*_{BLIP}(s\tau)^{1/2}. \tag{5.17}$$

Based on the above considerations, long lifetimes are required to achieve large gains in the SNR. A useful improvement in detectivity relative to a BLIP-limited discrete device can be achieved when the value of $s\tau$ exceeds unity. The performance of the device can be described in terms of the number of BLIP-limited elements in a serial array giving the same SNR,

$$N_{eq}(BLIP) = 2s\tau. \tag{5.18}$$

The SPRITE detectors are fabricated from lightly doped ($\approx 5 \times 10^{14}$ cm^{-3}) $Hg_{1-x}Cd_xTe$. Both bulk material and epilayers are used.[85] Single-, 2-, 4-, 8-, 16-, and 24-element arrays have been demonstrated; the 8-element arrays are the most common at present (Fig. 5.17). In order to manufacture the devices along a line, it is necessary to reduce the width of the readout zone and corresponding contacts to bring the contacts parallel to the length of the element within the width of the element. Various modifications of the device geometry have been proposed to improve both the detectivity and the spatial resolution, including horn geometry of the readout zone (Fig. 5.18) to reduce the transit time spread, and a slight tapering of the drift region to compensate for a slight change of drift velocity due to background radiation.

As shown in Table 5.1, to achieve usable performance in the 8–14-μm band, the SPRITE devices require liquid nitrogen (LN) cooling while three- or four-stage Peltier coolers are sufficient for effective operation in the 3–5 μm range.

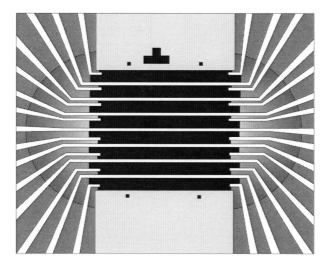

Figure 5.17 Schematic of an 8-element SPRITE array with bifurcated readout zones. (Reprinted from Ref. 82 with permission from Elsevier.)

Figure 5.18 Schematic geometry of drift/readout regions of SPRITE detectors: (a) straight/bifurcated, (b) straight/horn, and (c) tapered/horn. (Reprinted from Ref. 7 with permission from Elsevier.)

Table 5.1 Performance of SPRITE detectors. (Reprinted from Ref. 82 with permission from Elsevier.)

Material	Mercury cadmium telluride	
Number of elements	8	
Filament length (μm)	700	
Nominal sensitive area (μm)	62.5×62.5	
Operating band (μm)	8–14	3–5
Operating temperatures (K)	77	190
Cooling method	Joule-Thompson or heat engine	Thermoelectric
Bias field (V/cm)	30	30
Field of view	$f/2.5$	$f/2.0$
Ambipolar mobility (cm^2/Vs)	390	140
Pixel rate per element (pixel/s)	1.8×10^6	7×10^5
Typical element resistance (Ω)	500	4.5×10^3
Power dissipation (mW per element)	9	1
total	< 80	< 10
Mean D^* (500 K, 20 kHz, 1 Hz) 62.5×62.5 μm (10^{10} cm Hz$^{1/2}$/W)	> 11	4–7
Responsivity (500 K), 62.5×62.5 μm (10^4 V/W)	6	1

The response of the SPRITE detector to high spatial frequencies in an image is limited fundamentally either by spatial averaging due to the limited size of the read-out zone, or by the diffusive spread of photogenerated carriers in the filament.[83] The response can be further degraded by imperfect matching of the carrier drift velocity to the image velocity. The resolution size for 8–14-μm devices is $\approx$ 55 μm. For 200 K 3–5-μm devices, the spatial resolution is $\approx$ 140 μm.

One possible method to improve the resolution is to use a short device where the transit time is less than the lifetime to reduce the diffusive spread. The spatial resolution also can be improved by using anamorphic optics,[87] which increases the magnification of the image in the scan direction.[87] The detector length and scan speed are increased in the same proportion as the increase in magnification, but the diffusion length remains constant so that the spatial resolution is improved. Since SPRITE detectors remain background-limited even in low background flux, the SNR remains unaffected.

A number of improvements have been made to the spatial and thermal resolutions.[85] System thermal sensitivity can be increased by the use of a larger number of elements. In addition to parallel arrays, 2D 8×4 parallel/serial arrays have been demonstrated. Moving from an operating temperature of ≈ 80 K to ≈ 70 K with the cutoff wavelength shift from 12 to 12.5 μm can improve both signal-to-noise performance and spatial resolution. In the 3–5-μm band, the main improvement has been achieved by using more effective five- and six-stage Peltier coolers; 500 K blackbody detectivities as high as 5×10^{10} cm Hz$^{1/2}$/W have been achieved with 8-element arrays.

The ultimate size of SPRITE arrays is limited by the significant heat load imposed by Joule heating. It is expected that the equivalent performance of an array with more than 1000 elements in the 8–14-μm range will be achieved with a heat load less than 1 W.[88] Arrays equivalent to hundreds of conventional elements should be possible in the 3–5-μm band within the constraints imposed by Peltier coolers.

Despite remarkable successes, SPRITE detectors have important limitations such as limited size, stringent cooling requirements, and the necessity to use fast mechanical scanning. This means that SPRITE detectors are transition-stage devices to the staring 2D arrays era.

References

1. W. D. Lawson, S. Nielsen, E. H. Putley, and A. S. Young, "Preparation and properties of HgTe-CdTe," *J. Phys. Chem. Solids* **9**, 325–329 (1959).
2. B. E. Bartlett, D. E. Charlton, W. E. Dunn, P. C. Ellen, M. D. Jenner, and M. H. Jervis, "Background limited photoconductive detectors for use in the 8–14 micron atmospheric window," *Infrared Phys.* **9**, 35–36 (1969).
3. M. A. Kinch and S. R. Borrello, "0.1 eV HgCdTe photodetectors," *Infrared Phys.* **15**, 111–124 (1975).
4. M. A. Kinch, S. R. Borrello, and A. Simmons, "0.1 eV HgCdTe photoconductive detector performance," *Infrared Phys.* **17**, 127–135 (1977).
5. M. A. Kinch, S. R. Borrello, D. H. Breazeale, and A. Simmons, "Geometrical enhancement of HgCdTe photoconductive detectors," *Infrared Phys.* **17**, 137–145 (1977).
6. S. R. Borrello, M. Kinch, and D. Lamont, "Photoconductive HgCdTe detector performance with background variations," *Infrared Phys.* **17**, 121–125 (1977).
7. C. T. Elliott and N. T. Gordon, "Infrared detectors," in *Handbook on Semiconductors*, Vol. 4, C. Hilsum (Ed.), North-Holland, Amsterdam, 841–936 (1993).

8. S. P. Emmons and K. L. Ashley, "Minority-carrier sweepout in 0.09-eV HgCdTe," *Appl. Phys. Lett.* **20**, 241–242 (1972).

9. M. R. Johnson, "Sweep-out effects in $Hg_{1-x}Cd_xTe$ photoconductors," *J. Appl. Phys.* **43**, 3090–3093 (1972).

10. Y. J. Shacham-Diamond and I. Kidron, "Contact and bulk effects in intrinsic photoconductive infrared detectors," *Infrared Phys.* **21**, 105–115 (1981).

11. T. Ashley and C. T. Elliott, "Accumulation effects at contacts to n-type cadmium-mercury-telluride photoconductors," *Infrared Phys.* **22**, 367–376 (1982).

12. M. White, "Recombination in a graded n-n contact region in a narrow-gap semiconductor," *J. Phys. C: Solid State Phys.* **17**, 4889–4896 (1984).

13. D. L. Smith, "Effect of blocking contacts on generation-recombination noise and responsivity in intrinsic photoconductors," *J. Appl. Phys.* **56**, 1663–1669 (1984).

14. D. K. Arch, R. A. Wood, and D. L. Smith, "High responsivity HgCdTe heterojunction photoconductor," *J. Appl. Phys.* **58**, 2360–2370 (1985).

15. C. A. Musca, J. F. Siliquini, B. D. Nenner, and L. Faraone, "Enhanced performance of HgCdTe LWIR photoconductors passivated with higher bandgap HgCdTe," *COMAD '94 Proc.*, V. W. L. Chin and T. L. Tansley (Eds.), Sydney, Australia, 81–86 (1994).

16. C. A. Musca, J. E. P. Smith, J. M. Dell, and L. Faraone, "Performance and stability of HgCdTe photoconductive devices: A study of passivation and contact technology," *COMAD '98 Proc.*, UWA, Perth, Australia, MP-31–MP-32 (1998).

17. J. Piotrowski, W. Galus, and M. Grudzień, "Near room-temperature IR photodetectors," *Infrared Phys.* **31**, 1–48 (1990).

18. J. Piotrowski, "Uncooled operation of IR photodetectors," *Opto-Electr. Rev.* **12**, 111–122 (2004).

19. J. Piotrowski and A. Rogalski, "Uncooled long wavelenth infrared photon detectors," *Infrared Phys. Technol.* **46**, 115–131 (2004).

20. J. Piotrowski, Z. Orman, Z. Nowak, J. Pawluczyk, J. Pietrzak, A. Piotrowski, and D. Szabra, "Uncooled long wavelength infrared photodetectors with optimized spectral response at selected spectral ranges," *Proc. SPIE* **5783**, 616–624 (2005).

21. T. Ashley and C. T. Elliott, "Nonequilibrium devices for infra-red detection," *Electron. Lett.* **21**, 451–452 (1985).

22. T. Ashley, C. T. Elliott, and A. M. White, "Non-equilibrium devices for infrared detection," *Proc. SPIE* **572**, 123–132 (1985).

23. J. Auth, D. Genzow, and K. H. Herrmann. *Photoelectrische Erscheinungen*, Akademie-Verlag, Berlin (1977).

24. R. M. Broudy and V. J. Mazurczyk, "$Hg_{1-x}Cd_xTe$ photoconductors," in *Semiconductors and Semimetals*, Vol. 18, R. K. Willardson and A. C. Beer (Eds.), Academic Press, New York, 157–199 (1981).

25. Z. Wei-jiann and Z. Xin-Chen, "Experimental studies on low frequency noise of photoconductors," *Infrared Phys.* **33**, 27–31 (1992).

26. W. Galus, T. Persak, and J. Piotrowski. "Measurement of the $\alpha\tau/n_i$ factor determining maximum responsivity of the uncooled photoconductive CdHgTe detector operating at 10.6 μm wavelength," in *Proc. 8th Int. Symp. of the Technical Committee on Photon Detectors*, edited by IMEKO Secretariat, Budapest, 387–394 (1978).

27. J. F. Siliquini, C. A. Musca, B. D. Nener, and L. Faraone, "Performance of optimized $Hg_{1-x}Cd_xTe$ long wavelength infrared photoconductors," *Infrared Phys. Technol.* **35**, 661–671 (1994).

28. J. F. Siliquini, K. A. Fynn, B. D. Nener, L. Faraone, and R. H. Hartley, "Improved device technology for epitaxial $Hg_{1-x}Cd_xTe$ infrared photoconductor arrays," *Semicond. Sci. Technol.* **9**, 1515–1522 (1994).

29. J. F. Siliquini, C. A. Musca, B. D. Nener, and L. Faraone, "Temperature dependence of $Hg_{0.68}Cd_{0.32}Te$ photoconductor performance," *IEEE Trans. Electron Devices* **42**, 1441–1448 (1995).

30. C. A. Musca, J. F. Siliquini, K. A. Fynn, B. D. Nener, S. J. C. Irvine, and L. Faraone, "MOCVD grown wider-bandgap capping layers in $Hg_{1-x}Cd_xTe$ long-wavelength infrared photoconductors," *Semicond. Sci. Technol.* **11**, 1912–1922 (1996).

31. C. A. Musca, J. F. Siliquini, B. D. Nener, L. Faraone, and R. H. Hartley, "Enhanced responsivity of HgCdTe infrared photoconductors using MBE grown heterostructures," *Infrared Phys. Technol.* **38**, 163–167 (1997).

32. C. A. Musca, J. F. Siliquini, B. D. Nener, and L. Faraone, "Heterojunction blocking contacts in MOCVD grown $Hg_{1-x}Cd_xTe$ long wavelength infrared photoconductors," *IEEE Trans. Electron Devices* **44**, 239–249 (1997).

33. G. Parish, C. A. Musca, J. F. Siliquini, J. Antoszewski, J. M. Dell, B. D. Nener, L. Faraone, and G. J. Gouws, "A monolithic dual-band HgCdTe infrared detector structure," *IEEE Electron Device Letters* **18**, 352–354 (1997).

34. M. Małachowski, A. Józwikowska, K. Józwikowski, and J. Piotrowski, "Influence of contacts on infrared photoconductor performance," *Proc. SPIE* **1845**, 71–76 (1992).

35. J. R. Lowney, D. G. Seiler, W. R. Thurber, Z. Yu, X. N. Song, and C. L. Littler, "Heavily accumulated surfaces of mercury cadmium telluride detectors: Theory and experiment," *J. Electron. Mater.* **22**, 985–991 (1993).

36. M. White, "Generation-recombination processes and Auger suppression in small-bandgap detectors," *J. Cryst. Growth* **86**, 840–848 (1988).

37. T. Elliott, "Non-equilibrium modes of operation of narrow-gap semiconductor devices," *Semicond. Sci. Technol.* **5**, S30–S37 (1990).

38. T. Ashley, C. T. Elliott, N. T. Gordon, R. S. Hall, A. D. Johnson, and G. J. Pryce, "Negative luminescence from $In_{1-x}Al_xSb$ and $Cd_xHg_{1-x}Te$ diodes," *Infrared Phys. Technol.* **36**, 1037–1044 (1995).

39. T. Ashley, C. T. Elliott, N. T. Gordon, R. S. Hall, A. D. Johnson, and G. J. Pryce, "Room temperature narrow gap semiconductor diodes as sources

and detectors in the 5–10 μm wavelength region," *J. Crystal Growth* **159**, 1100–1103 (1996).

40. A. Józwikowska, "Application of nonequilibrium phenomena in $Hg_{1-x}Cd_xTe$ photodetectors," thesis, Wojskowa Akademia Techniczna (WAT), Warsaw (1992).

41. J. Piotrowski, A. Józwikowska, K. Józwikowski, and R. Ciupa, "Numerical analysis of long wavelength extracted photodiodes," *Infrared Phys.* **34**, 565–572 (1993).

42. T. Ashley, C. T. Elliott, and A. T. Haker, "Non-equilibrium mode of operation for infrared detection," *Infrared Phys.* **26**, 303–315 (1986).

43. J. Piotrowski, "$Hg_{1-x}Cd_xTe$ Infrared Photodetectors," in *Infrared Photon Detectors*, 391–494, SPIE Press, Bellingham, WA (1995).

44. D. L. Spears, "Heterodyne and direct detection at 10.6 μm with high temperature p-type photoconductors," *Proc. IRIS Active Systems*, San Diego, 1–15 (1982).

45. D. Long and J. L. Schmit, "Mercury-Cadmium Telluride and Closely Related Alloys," in *Semiconductors and Semimetals*, Vol. 5, R. K. Willardson and A. C. Beer (Eds.), Academic Press, New York, 175–255 (1970).

46. Z. Djuric, "Isothermal vapour-phase epitaxy of mercury-cadmium telluride (Hg,Cd)Te," *J. Materials Science: Materials in Electronics* **5**, 187–218 (1995).

47. K. Nagahama, R. Ohkata, and T. Murotani, "Preparation of high quality n-$Hg_{0.8}Cd_{0.2}Te$ epitaxial layer and its application to infrared detector ($\lambda = $ 8–14 μm)," *J. J. Appl. Phys.* **21**, L764–L766 (1982).

48. T. Nguyen-Duy and D. Lorans, "Highlights of recent results on HgCdTe thin film photoconductors," *Semicond. Sci. Technol.* **6**, 93–95 (1991)

49. B. Doll, M. Bruder, J. Wendler, J. Ziegler, and H. Maier, "3–5 μm photoconductive detectors on liquid-phase-epitaxial-MCT," in *Fourth Int. Conf. Advanced Infrared Detectors and Systems*, IEE, London, 120–124 (1990).

50. J. F. Siliquini, C. A. Musca, B. D. Nener, and L. Faraone, "Performance of optimized $Hg_{1-x}Cd_xTe$ long wavelength infrared photoconductors," *Infrared Phys. Technol.* **35**, 661–671 (1994).

51. L. T. Specht, W. E. Hoke, S. Oguz, P. J. Lemonias, V. G. Kreismanis, and R. Korenstein, "High performance HgCdTe photoconductive devices grown by metalorganic chemical vapor deposition," *Appl. Phys. Lett.* **48**, 417–418 (1986).

52. C. G. Bethea, B. F. Levine, P. Y. Lu, L. M. Williams, and M. H. Ross, "Photoconductive $Hg_{1-x}Cd_xTe$ detectors grown by low-temperature metalorganic chemical vapor deposition," *Appl. Phys. Lett.* **53**, 1629–1631 (1988).

53. R. Druilhe, A. Katty, and R. Triboulet, "MOVPE grown (Hg,Cd)Te layers for room temperature operating 3–5 μm photoconductive detectors," in *Fourth Int. Conf. Advanced Infrared Detectors and Systems*, 20–24, IEE, London (1990).

54. M. C. Chen and M. J. Bevan, "Room-temperature midwavelength two-color infrared detectors with HgCdTe multilayer structures by metal-organic chemical-vapor deposition," *J. Appl. Phys.* **78**, 4787–4789 (1995).

55. S. Yuan, J. Yu, M. Yu, Y. Qiao, and J. Zhu, "Infrared photoconductor fabricated with a molecular beam epitaxially grown CdTe/HgCdTe heterostructure," *Appl. Phys. Lett.* **58**, 914–916 (1991).

56. Y. Nemirovsky and G. Bahir, "Passivation of mercury cadmium telluride surfaces," *Vac. Sci. Technol. A* **7**, 450–459 (1989).

57. P. C. Catagnus and C. T. Baker, "Passivation of Mercury Cadmium Telluride," U. S. Patent 3,977,018 (1976).

58. Y. Nemirovsky and E. Finkman, "Anodic oxide films on $Hg_{1-x}Cd_xTe$," *J. Electrochem. Soc.* **126**, 768–770 (1979).

59. E. Bertagnolli, "Improvement of anodically grown native oxides on n-(Cd,Hg)Te," *Thin Solid Films* **135**, 267–275 (1986).

60. A. Gauthier, "Process for Passivation of Photoconductive Detectors Made of HgCdTe," U. S. Patent 4,624,715 (1986).

61. Y. Nemirovsky and R. Goshen, "Plasma anodization of $Hg_{1-x}Cd_xTe$," *Appl. Phys. Lett.* **37**, 813–814 (1980).

62. S. P. Buchner, G. D. Davis, and N. E. Byer, "Summary Abstract: Photochemical oxidation of (Hg,Cd)Te," *J. Vac. Sci. Technol.* **21**, 446–447 (1982).

63. Y. Shacham-Diamand, T. Chuh and W. G. Oldham, "The electrical properties of Hg-sensitized 'Photox'-oxide layers deposited at 80°C," *Solid-State Electron.* **30**, 227–233 (1987).

64. M. V. Blackman, D. E. Charlton, M. D. Jenner, D. R. Purdy, and J. T. M. Wotherspoon, "Type conversion in $Hg_{1-x}Cd_xTe$ by ion beam treatment," *Electron. Lett.* **23**, 978–979 (1987).

65. P. Brogowski, H. Mucha, and J. Piotrowski, "Modification of mercury cadmium telluride, mercury manganese tellurium, and mercury zinc telluride by ion etching," *Phys. Stat. Sol. (a)* **114**, K37–K40 (1989).

66. P. H. Zimmermann, M. B. Reine, K. Spignese, K. Maschhoff, and J. Schirripa, "Surface passivation of HgCdTe photodiodes," *J. Vac. Sci. Technol. A* **8**, 1182–1184 (1990).

67. Y. Nemirovsky, L. Burstein, and I. Kidron, "Interface of p-type $Hg_{1-x}Cd_xTe$ passivated with native sulfides," *J. Appl. Phys.* **58**, 366–373 (1985).

68. E. Weiss and C. R. Helms, "Composition, growth mechanism, and stability of anodic fluoride films on $Hg_{1-x}Cd_xTe$," *J. Vac. Sci. Technol. B* **9**, 1879–1885 (1991).

69. Y. Nemirovsky, "Passivation with II-VI compounds," *J. Vac. Sci. Technol. A* **8**, 1185–1187 (1990).

70. G. Sarusi, G. Cinader, A. Zemel, and D. Eger, "Application of CdTe epitaxial layers for passivation of p-type $Hg_{0.77}Cd_{0.23}Te$," *J. Appl. Phys.* **71**, 5070–5076 (1992).

71. A. Mestechkin, D. L. Lee, B. T. Cunningham, and B. D. MacLeod, "Bake stability of long-wavelength infrared HgCdTe photodiodes," *J. Electron. Mater.* **24**, 1183–1187 (1995).

72. G. Bahir, V. Ariel, V. Garber, D. Rosenfeld, and A. Sher, "Electrical properties of epitaxially grown CdTe passivation for long-wavelength HgCdTe photodiodes," *Appl. Phys. Lett.* **65**, 2725–2727 (1994).

73. J. E. Slavek and H. H. Randal, "Optical immersion of HgCdTe photoconductive detectors," *Infrared Phys.* **15**, 339–340 (1975).

74. I. C. Carmichael, A. B. Dean, and D. J. Wilson, "Optical immersion of a cryogenically cooled 77 K photoconductive CdHgTe detector," in *Second Int. Conf. on Advanced Infrared Detectors and Systems*, 45–48, IEE, London (1983).

75. M. Grudzien and J. Piotrowski, "Monolithic optically immersed HgCdTe IR detectors," *Infrared Phys.* **29**, 251–253 (1989).

76. N. T. Gordon, "Design of $Hg_{1-x}Cd_xTe$ infrared detector arrays using optical immersion with microlenses to achieve a higher operation temperature," *Semicond. Sci. Technol.* **8**, C106–C109 (1991).

77. J. Piotrowski, M. Grudzień, Z. Nowak, Z. Orman, J. Pawluczyk, M. Romanis, and W. Gawron, "Uncooled photovoltaic $Hg_{1-x}Cd_xTe$ LWIR detectors," *Proc. SPIE* **4130**, 175–184 (2000).

78. Data sheets of these companies: Hamamatsu, Judson Technologies (www.judsontechnologies.com), and Vigo System S.A. (www.vigo.com.pl).

79. H. Halpert and B. L. Musicant, "N-Color (Hg,Cd)Te photodetectors," *Appl. Opt.* **11**, 2157–2161 (1972).

80. Y. Li and D. J. Adams, "Experimental studies on the performance of $Hg_{0.8}Cd_{0.2}Te$ photoconductors operating near 200 K," *Infrared Phys.* **35**, 593–595 (1994).

81. C. T. Elliott, D. Day, and D.J. Wilson, "An integrating detector for serial scan thermal imaging," *Infrared Phys.* **22**, 31–42 (1982).

82. A. Blackburn, M. V. Blackman, D. E. Charlton, W. A. E. Dunn, M. D. Jenner, K. J. Oliver, and J. T. M. Wotherspoon, "The practical realization and performance of SPRITE detectors," *Infrared Phys.* **22**, 57–64 (1982).

83. T. Ashley, C. T. Elliott, A .M. White, J. T .M. Wotherspoon, and M. D. Johns, "Optimization of spatial resolution in SPRITE detectors," *Infrared Phys.* **24**, 25–33 (1984).

84. J. T. M. Wotherspoon, R. J. Dean, M. D. Johns, T. Ashley, C. T. Elliott, and M. A. White, "Developments in SPRITE infra-red detectors," *Proc. SPIE* **810**, 102–112 (1984).

85. J. Severn, D. A. Hibbert, R. Mistry, C. T. Elliott, and A. P. Davis, "The design and performance options for SPRITE arrays," in *Fourth Int. Conf. Advanced Infrared Detectors and Systems*, IEE, London, 9–14 (1990).

86. A. P. Davis, "Effect of high signal photon fluxes on the responsivity of SPRITE detectors," *Infrared Phys.* **33**, 301–305 (1992).

87. A. Campbell, C. T. Elliott, and A. M. White, "Optimisation of SPRITE detectors in anamorphic imaging systems," *Infrared Phys.* **27**, 125–133 (1987).

88. C. T. Elliott, "Infrared detectors with integrated signal processing," in *Solid State Devices*, A. Goetzberger and M. Zerbst (Eds.), Verlag Chemie, Weinheim, 175–201 (1983).

Chapter 6

$\mathrm{Hg}_{1-x}\mathrm{Cd}_x\mathrm{Te}$ Photodiodes

The photovoltaic (PV) effects in $\mathrm{Hg}_{1-x}\mathrm{Cd}_x\mathrm{Te}$ were mentioned for the first time in 1959 by Lawson et al.[1] The development of $\mathrm{Hg}_{1-x}\mathrm{Cd}_x\mathrm{Te}$ photodiodes was stimulated initially by their applications as high-speed detectors, mostly for direct and heterodyne detection of 10.6-μm CO_2 laser radiation. Later efforts have been dominated by applications for large arrays used in LWIR and MWIR thermal imaging systems. In contrast to photoconductors, photodiodes have a very low power dissipation and can be assembled in 2D arrays containing a very large ($>10^6$) number of elements, limited only by existing technologies rather than by heat dissipation issues. Photodiodes are the most promising devices for uncooled operation, and at present, significant efforts are on Auger-suppressed devices[2] and multiple heterojunctions.[3]

$\mathrm{Hg}_{1-x}\mathrm{Cd}_x\mathrm{Te}$ photodiodes have been reviewed by several authors.[4–7] This chapter will concentrate on the specific features of photodiodes pertinent to the uncooled operation of the devices.

6.1 Theoretical Design of $\mathrm{Hg}_{1-x}\mathrm{Cd}_x\mathrm{Te}$ Photodiodes

6.1.1 Photocurrent and dark current of photodiodes

Generally speaking, a photodiode is a semiconductor structure containing an absorber region with one or two contacts that sense the photogenerated carriers. But in contrast to a photoconductor, a photodiode can be operated at zero or at reverse bias. The contacts can be homojunctions or heterojunctions. For the best performance, the contacts must not introduce minority carriers to the absorber. At the same time, the contacts should effectively collect the charge carriers generated in the absorber, so the contacts must not form barriers for minority carriers.[8–12] This can be accomplished by using wide-gap and heavily doped heterojunction contacts. Heavily-doped, band-filled n^+-$\mathrm{Hg}_{1-x}\mathrm{Cd}_x\mathrm{Te}$ is also a near ideal contact for p-type absorbers; as a result, a heterojunction contact may not be necessary. However, since the band-filling effect does not occur in p-type $\mathrm{Hg}_{1-x}\mathrm{Cd}_x\mathrm{Te}$, the only choice in that case is a wide-gap p-type contact to an n-type absorber.

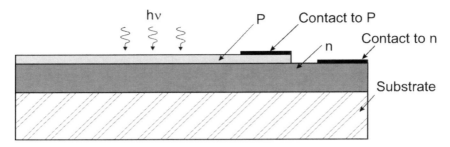

Figure 6.1 Schematic cross-section of a double-layer heterojunction device with top contact on the p-type layer and a remote ohmic contact to the n-layer.

Consider as an example a double-layer heterojunction photodiode with a narrow-gap n-type absorber region, and a wide-gap and p-type cap (as shown in Fig. 6.1). The device can be front-side illuminated through a wide-gap p-type layer or it can be back-side illuminated. The photocurrent is due to the optical generation of charge carriers in absorber regions that diffuse to form a junction where they are collected. For efficient absorption of incoming radiation, the thickness of the absorber region must be comparable to the absorption depth. At the same time, the junction should be located at a distance less than the diffusion length from the generation site. As a result, the diffusion length should be larger than the absorption length. This relation between diffusion length and absorption length is readily achieved in cooled photodiodes where the diffusion length is typically much larger than the absorption length. In contrast, the absorption depth in narrow-bandgap semiconductors that are operating close to room temperature may be well in excess of the diffusion length for any gap and doping. Consequently, IR radiation is only partial absorped in the region from which the carriers can be collected. This precludes the possibility of obtaining near-unity quantum efficiency in such a device for any thickness of the absorber region. That problem can be solved by using multiple heterojunction devices.[13]

The dark current is due to thermal generation of charge carriers. All regions of the photodiode may contribute to the resulting dark current.[4–7] In practice, nonfundamental sources dominate the dark current of the present 80 K $Hg_{1-x}Cd_xTe$ LWIR and MWIR photodiodes. In contrast, the dark current of properly designed photodiodes that are operating at room temperature is frequently determined by thermal generation due to fundamental mechanisms. Therefore, most of the efforts for uncooled detectors are directed toward the development of the specific photodiode structure in which the fundamental mechanisms can be suppressed. Table 6.1 shows the main sources of dark current in photodiodes.

6.1.2 Structures of $Hg_{1-x}Cd_xTe$ photodiodes

The basic steps for developing $Hg_{1-x}Cd_xTe$ photodiode structures are n-p and n^+-p homojunction, double-layer P-n heterojunction, and three or more layer het-

Table 6.1 The main sources of a photodiode dark current.

Mechanism	Methods to Minimize Dark Current
Absorber	
Auger generation	• optimized doping in photodiodes operating at equilibrium mode
	• suppressed to zero by nonequilibrium depletion
internal radiative generation	• optimized doping
external radiative generation	• optical shielding from radiation coming from passive region of device and from other elements in an array
	• optical concentrators, e.g. immersion lenses
Shockley-Read generation	• improved material technology
$1/f$ noise	• improved technology
Depletion Layer	
generation-recombination	• reduction of concentration of SR centers by improved material technology
	• optimized doping
	• reduction of size of junction compared the pixel size
band-to-band tunneling	• reduced doping
	• heterojunction
trap-assisted tunneling	• improved material quality
avalanche current	• reduced doping
	• heterojunction
leakage due to dislocations	• reduced dislocation density
leakage due to precipitates	• improved material technology
$1/f$ noise	• improved technology
	• reduced dislocation density
Cap Layer	
Auger generation in the bulk	• optimized band gap and doping
Auger generation at graded interface	• optimized grading of bandgap and doping
Shockley-Read generation	• reduction of concentration of SR centers by improved material technology
Surfaces	
generation at surface states	• improved passivation
	• heterojunction passivation
shunt leakage due to inversion layer	• improved passivation
	• optimized acceptor doping
tunneling induced near the surface	• improved passivation
	• optimized acceptor doping
avalanche multiplication in a field-induced surface region	• improved passivation
	• optimized acceptor doping
Contacts	
generation at contacts	• wide gap or/and degenerately doped cap material
	• remote contacts
Johnson-Nyquist noise of parasitic resistance	• improved design

erostructures. This is illustrated in Fig. 6.2, which shows schematic cross-sections of the devices. The architecture of practical devices is more complex, as will be discussed below.

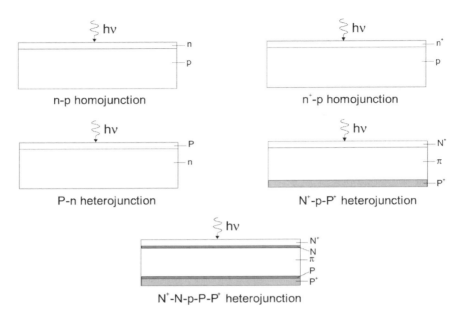

Figure 6.2 Schematic cross-sections of principal structures of $Hg_{1-x}Cd_x Te$ photodiodes: n-p and n^+-p homojunctions, P-n heterojunction, N^+-p-P^+, and N^+-N-p-P-P^+ heterostructures.

Homojunction n-p device

Homojunction n-p devices were the first practical $Hg_{1-x}Cd_x Te$ photodiodes. Both the n- and p-regions contribute to the resulting quantum efficiency, dark current, and noise. The device is difficult to optimize. The illuminated diode region should be thin enough ($<L_D$) to ensure effective collection of photogenerated carriers by junction. The other layer also should be thin enough to reduce the dark current caused by bulk thermal generation in that layer. But when the layers are thin, the device will be vulnerable to the surface generation-recombination processes. This is especially pronounced when the surfaces are metalized for contacts.

Homojunction n^+-p device

Problems with the surface processes can be solved with a degenerately doped n^+ layer. Due to the pronounced Burnstein-Moss effect, the n^+-layer does not contribute to the resulting quantum efficiency or dark current generation. The n^+-layer plays the roles of a window layer and a contact that collects minority carriers. The p-region plays the role of an absorber, and ideally remains the only source of dark current. In practice, the device may be vulnerable to surface processes at the p-type region surface. The remote ohmic contact to p-layers must be used.

Double-layer heterojunction P$^+$-n photodiode

The n-layer is the absorber, while the wide-gap and heavily doped P-layer forms the junction and may play the role of window layer. The remote ohmic contact to the n-layer must be used.

Triple-layer heterojunction (N$^+$-n-P$^+$, N$^+$-p-P$^+$, n$^+$-n-P$^+$, n$^+$-p-P$^+$) photodiodes

Narrow-gap n, p, ν, or π layers act as the absorber, while wide-gap and heavy-doped N$^+$ and P$^+$ layers form junctions and may play the role of window layer. These devices are better protected against dark current generation at surfaces, contacts, and interfaces. In addition, heavy doping makes it possible to achieve low-resistance contacts. A heavy doped n$^+$-layer is sometimes used instead of the N$^+$-layer.

Four- and five-layer (N$^+$-n-P-P$^+$, N$^+$-p-P-P$^+$, N$^+$-N-n-P-P$^+$, N$^+$-N-p-P-P$^+$) heterojunction photodiodes

Additional wide-gap layers but fewer doped-gap layers are frequently placed on purpose between the absorber and the heavy-doped regions to prevent thermal generation at interfaces and tunnel currents.[14]

With the exception of the p-n homojunction photodiode, more advanced photodiode structures have a one-layer absorber region of n- or p-type conductivity, which ideally is the only source of dark current.

Consider a single absorber device. We will assume that the diffusion length in the absorber layer is larger than the absorption depth. For the best performance, the doping of the absorber should produce a hole concentration that results in the lowest possible thermal generation rate. The thickness of the absorber should be optimized for the best compromise between the requirements of high quantum efficiency and a low dark current. This is achieved with a base thickness close to $d = 1.26/\alpha$ or $0.63/\alpha$ for 0 and 1 backside reflection coefficients (see Chapter 2).[13–16]

The current-voltage characteristic of the ideal photodiode is described as

$$I = I_{ph} + I_s \left[\exp\left(\frac{qV}{kT}\right) - 1 \right]. \tag{6.1}$$

The short circuit photocurrent I_{ph} is

$$I_{ph} = R_i \Phi, \tag{6.2}$$

where Φ is the radiation power and R_i is the current responsivity:

$$R_i = \frac{\lambda \eta q}{hc}. \tag{6.3}$$

Assuming that the dark current saturation is only due to thermal generation in the absorber, the unity photoelectric gain, and the absorber is thin relative to the diffusion length,

$$J_s = Gtq, \tag{6.4}$$

where G is the generation rate in the base layer.

It should be noted here that the assumptions of constant thermal generation and unity photoelectric gain might not be valid for $Hg_{1-x}Cd_xTe$ photodiodes where thermal generation and recombination are governed by the Auger processes. Nonetheless, this simple calculation makes it possible to correctly predict the ultimate performance of $Hg_{1-x}Cd_xTe$ photodiodes operating at zero bias (with the exception of junctions with a near-intrinsic absorber[15-17]).

Figure 6.3 shows the calculated saturation currents as a function of wavelength and temperature. The calculations have been performed for the optimized doping close to that resulting in a $p = \gamma^{1/2}n_i$ hole concentration. The thickness of $0.63/\alpha$ has been assumed as the optimum thickness for the double radiation pass device. The saturation dark-current densities are very large compared to those of LN-cooled devices. For example, J_s for an uncooled photodiode that is optimized for $\lambda \approx 10$ μm exceeds 1000 A/cm^2 and is by four orders of magnitude larger than the photocurrent due to the 300 K background radiation. In contrast, the dark current of a 5-μm device operating at 230 K is equal to the background radiation generated by the photocurrent.

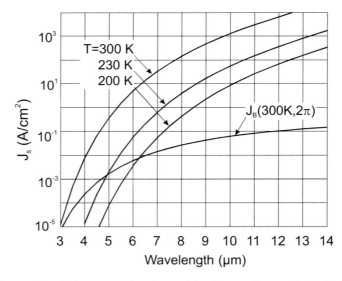

Figure 6.3 Saturation dark-current density and background-generated photocurrent as a function of wavelength.

The noise current that is due to the thermal generation-recombination processes in the absorber and the background optical generation is

$$I_n^2 = 2\big[(G + R) + \eta \Phi_B\big] q^2 A t \Delta f. \tag{6.5}$$

Photodiodes are typically operated at zero bias to minimize the heat load and to ensure zero $1/f$ noise. At zero bias, the thermal generation and recombination rates are equal. In this case,

$$I_n^2 = 2\big(2G + q^2 \eta \Phi_B\big) q^2 A t \Delta f, \tag{6.6}$$

or

$$I_n^2 = (4J_s + 2J_B) A t \Delta f. \tag{6.7}$$

It should be noted that the noise of an ideal photodiode operating at a reverse-bias recombination occurs at the neutral region of the device and does not contribute to the noise. In this case,

$$I_n^2 = 2(J_s + J_B) A \Delta f. \tag{6.8}$$

The normalized detectivity for zero-bias conditions is

$$D^* = \frac{\lambda \eta q}{hc} \frac{1}{(4J_s + 2J_B)^{1/2}}, \tag{6.9}$$

or, at reverse bias,

$$D^* = \frac{\lambda \eta q}{hc} \frac{1}{(2J_s + 2J_B)^{1/2}}. \tag{6.10}$$

The analysis of Eq. (6.10) and the results presented in Fig. 6.3 indicate that the detectivity of an uncooled 10-μm photodiode is approximately two orders of magnitude less than the BLIP performance, while the detectivity of a Peltier-cooled 5-μm device may approach the BLIP limit.

Frequently, the photodiode performance is discussed in terms of the zero bias resistance-area product. As follows for the ideal diode I-V plot,

$$R_o A = \frac{kT}{q J_s}, \tag{6.11}$$

or

$$R_o A = \frac{kT}{q^2 G t}. \tag{6.12}$$

The zero bias detectivity can be expressed as

$$D^* = \frac{\lambda \eta q}{hc} \frac{(R_o A)^{1/2}}{(4kT + 2q^2 \eta \Phi_B R_o A)^{1/2}}. \tag{6.13}$$

For the best performance under the given operating conditions (wavelength and temperature), the value of $\eta (R_o A)^{1/2}$ should be maximized. The $\eta (R_o A)^{1/2}$ is a photodiode figure of merit that determines the performance of a photodiode.

The $R_o A$ product is more frequently being used as the main photodiode figure of merit. It should be noted that $R_o A$ is not the sole figure of merit. In principle, a very large $R_o A$ can be obtained by making the thickness of the absorber low by sacrificing quantum efficiency and detectivity. Therefore, $R_o A$ can be used to characterize photodiode performance with the same quantum efficiency.

Compare $R_o A$ for p- and n-type absorbers of the same thickness. Assuming the Auger mechanism in the extrinsic p-type absorber,

$$R_o A = \frac{2kT\tau_{A7}^i}{q^2 N_a t}. \tag{6.14}$$

For the n-type absorber,

$$R_o A = \frac{2kT\tau_{A1}^i}{q^2 N_d t}. \tag{6.15}$$

As Eqs. (6.14) and (6.15) show, the $R_o A$ product can be increased by reducing the thickness of the base layer. As follows from the general considerations presented in Chapter 2, the optimum thickness is close to the absorption depth. Since $\gamma = \tau_{A7}^i / \tau_{A1}^i > 1$, higher $R_o A$ values can be achieved in p-type base devices than those of n-type devices with the same doping level. Maximum $R_o A$ is achievable with absorber doping producing $p = \gamma^{1/2} n_i$, which corresponds to the minimum thermal generation.[17] Therefore, optimum doping depends on the intrinsic concentration. The required level is quite high for temperatures that are above 200 K and wavelengths that are above 8 μm and within the range 1×10^{16} cm^{-3} to 5×10^{17} cm^{-3}. Under these conditions, the doping is easily achievable. In contrast, the MWIR devices that are operating with thermoelectric cooling may require an acceptor doping level below 1×10^{16} cm^{-3}, which may be difficult to obtain in practice. In some cases, slightly better performance can be obtained by using a lightly doped n-type absorber.

Figure 6.4 shows the calculated $R_o A$ product for p-type absorbers optimized for the best detectivity. The high-temperature photodiodes have quite low resistances. Extremely low resistance can be expected for an uncooled LWIR detector.

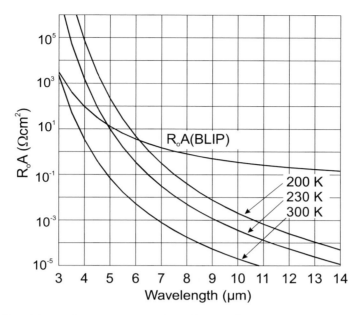

Figure 6.4 Zero bias resistance-area product of $Hg_{1-x}Cd_xTe$ photodiodes as a function of wavelength and temperature. The R_oA required when the 300 K thermal noise of a photodiode is equal to the background flux quantum noise is also plotted.

6.1.3 Current-voltage characteristics of photodiodes

The ideal $J = J_s \exp[(eV/kT) - 1]$ characteristic describes the situation when the photodiode current is caused by the thermal generation and recombination processes, the lifetime and generation rate of minority carriers on the low-doped side are unchanged by the changing value of the minority concentration in the vicinity of the junction, and the photoelectric gain is unity.

The assumptions are not more valid in $Hg_{1-x}Cd_xTe$ uncooled photodiodes. When the absorber is near intrinsic, thermal generation-recombination is governed by Auger mechanisms with important consequences to photodiode properties.[18] The properties of the devices are also strongly influenced by ambipolar effects due to a large electron-to-hole mobility ratio that may result in a significant gain in primary dark current.[19]

Consider first l-h, n^+-p, and p^+-n junctions with heavily and lightly doped semiconductor contacts that have a minority carrier concentrations in heavily doped regions so small that the minority carrier current is negligible and no surface generation at the junction interface occurs. Assuming that the length of the lightly doped zone is much longer than the ambipolar diffusion length, then

$$L_a = (D_a\tau)^{1/2}, \tag{6.16}$$

where D_a is the ambipolar diffusion coefficient

$$D_a = \frac{D_n D_p (n + p)}{D_n + D_p},\qquad (6.17)$$

and D_n and D_p are the diffusion coefficients for electrons and holes.

Under reverse bias, the minority carriers are extracted from the lightly doped regions at l-h junctions; electrons are extracted at the n^+-p junction and holes at the p^+-n junctions. The extracted carriers are partially replenished by the diffusion and drift transport from the bulk of the i-zone. The concentrations in the vicinity of the junctions will decrease, resulting in the partial suppression of Auger generation.

Consider now a three-layer n^+-n-p^+ or n^+-p-p^+ photodiode structure (Fig. 6.5) in which the length of the lightly doped zone is comparable to the ambipolar diffusion length so that the two junctions in the structure can interact. When the structure is reverse biased, the charge carriers are extracted from the lightly doped zone in the vicinity of the anisotype junctions. In contrast to the long, lightly doped zone device, the carriers cannot be replenished from the short zone or from heavily doped isotype contacts due to exclusion. At higher bias, if the length of the i-zone is short, this effect can be felt throughout the lightly doped zone, which results in a depletion of the majority carriers to the extrinsic level while the concentration of minority carriers can be reduced to a much lower level. Consequently, the suppression of Auger generation occurs with a dramatic decrease in dark current.[18] It is obvious that the high impedance device with a low dark current are of great interest for efficient detection of IR radiation.

An analytical model describes the carrier transport in idealized n^+-i, p^+-i, and n^+-i-p^+ structures with abrupt junctions, the total absence of doping in the intrinsic zone, the absence of other generation-recombination mechanisms, and the absence of nonparabolic effects in n^+-i and p^+-i structures.[19]

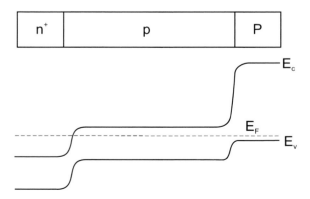

Figure 6.5 Three-layer n^+-p-P heterostructure showing energy levels at zero-bias conditions.

The J-V curves for n^+-i and p^+-i junctions can be expressed as

$$J_s = \frac{D_n q n_i}{L_a}\left[1 - \exp\left(-\frac{2q V_j}{kT}\right)\right] \quad \text{for } n^+\text{-i,} \tag{6.18}$$

and

$$J_s = \frac{D_p q n_i}{L_a}\left[1 - \exp\left(-\frac{2q V_j}{kT}\right)\right] \quad \text{for } p^+\text{-i,} \tag{6.19}$$

where V_j is the voltage applied to the junction. The total device voltage includes a significant contribution to the field in the i-zone. The saturation current for the n^+-i junction is larger than that for the p^+-i junction by a factor $D_n/D_p = \mu_n/\mu_p$.

Close examination of the last two equations reveals that the saturation currents correspond to halved thermal generation within the ambipolar diffusion length with unity gain for the p^+-i junction, and $D_n/D_p = \mu_n/\mu_p$ gain for the n^+-i junction. Correspondingly, the zero-bias resistance product of the junctions is

$$R_o A = \frac{1}{2}\frac{kT}{q J_s}. \tag{6.20}$$

Comparing Eq. (6.20) with the result for extrinsic diodes ($kT/q J_s$) shows that an Auger-dominated intrinsic junction has one-half the saturation current of an extrinsic diode with the same $R_o A$. The reason for the reduced saturation current is the quenching of the Auger generation as concentration decreases with reverse bias. A secondary effect is the modified local diffusion length induced by the locally varying lifetime.

The n^+-i junction is inherently more "soft" than the p^+-i junction with the difference in the resistance equal to the μ_n/μ_p ratio—almost two orders of magnitude. This anomalously low junction resistance is due to the ambipolar effects in the semiconductor with mixed conductivity and the large mobility ratio that results in μ_n/μ_p photoelectric gain in the vicinity of the n^+-i junction.

Consider now a three-layer n^+-i-p^+ photodiode structure (Fig. 6.6) in which the length of the i-zone is comparable to the ambipolar diffusion length, which allows the n^+-i and i-p^+ junctions to interact.

White reported an analytic solution for the ideal n^+-i-p^+ photodiode that predicts the reduction of leakage currents and negative resistance for n^+-i-p^+ structures.[19] Figure 6.6 shows a schematic distribution of the carrier concentration in the reverse-biased n^+-i-p^+ photodiode. The voltage applied to the device is a sum of voltages at n^+-i and i-p^+ junctions and voltage across the i-zone (see Fig. 6.7):

$$V = V_n + V_p + V_i. \tag{6.21}$$

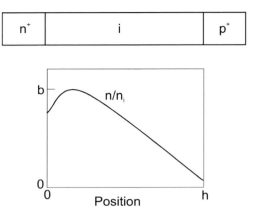

Figure 6.6 Distribution of carriers in the i-zone when μ_n larger than μ_p and for a device bias of $2kT/q$. The maximum of concentration approaches n^+-p junction with increasing μ_n/μ_p ratio. The maximum value of n/n_i is equal to parameter b.

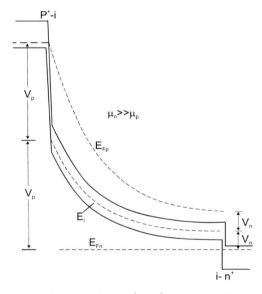

Figure 6.7 Schematic band diagram of the n^+-i-p^+ structure. (Reprinted from Ref. 19 with permission from Elsevier.)

The voltage across the i-zone is

$$V_i = \int E \, dx. \tag{6.22}$$

The electric field can be expressed in terms of the total current by

$$E = \frac{J}{q(\mu_n + \mu_p)} \frac{1}{n} + \frac{kT}{q} \frac{D_p - D_n}{D_p + D_n} \frac{1}{n} \frac{dn}{dx}. \tag{6.23}$$

The first term represents ohmic effects and is negligible in many cases. The second term is the Dember field, which integrates to

$$V_i = \frac{kT}{q}\frac{D_p - D_n}{D_p + D_n}\ln\frac{n_n}{n_p} = \frac{D_n - D_p}{D_p + D_n}(V_p - V_n). \qquad (6.24)$$

For $D_n \gg D_p$ and $V_n \ll V_p$, the Dember voltage equals the voltage drop at the p$^+$-i junction, so the total voltage of the device $V = 2V_p$. Therefore, the voltage drop across the i-zone due to the Dember effect cancels exactly the voltage drop across the "soft" n$^+$-i junction and doubles the voltage across the "hard" p$^+$-i junction. This effect is seen clearly in Fig. 6.7, which shows the quasi-Fermi levels and band bending.

Figure 6.8 shows the calculated I-V curves for different device lengths. For a long i-zone ($l \gg L_a$), the J-V characteristic is determined entirely by the voltage drop at the p$^+$-i junction and the Dember voltage nearby, resembling characteristics of conventional photodiodes:

$$J = -\frac{D_p q n_i}{L_a}\left[1 - \exp\left(\frac{qV}{kT}\right)\right]. \qquad (6.25)$$

The behavior of a device with its length comparable to L_a is more complex. At low bias, the device has low impedance, and the current rises sharply with voltage. At higher voltages, the current achieves its maximum value and starts to decrease,

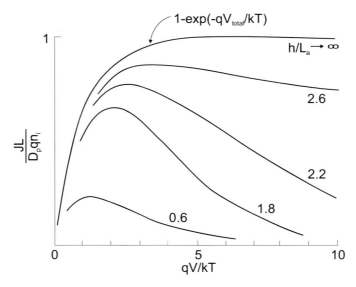

Figure 6.8 Current-voltage characteristics of the low-gap p-i-n diode structure for different i-zone lengths h (shown as the ratio h/L_a) and for $\mu_n \gg \mu_p$. For $h/L_a > \pi/2^{1/2} = 2.22$, the saturation current does not reduce to zero at very high bias voltage. (Reprinted from Ref. 19 with permission from Elsevier.)

producing negative resistance. If the length of the device is shorter than the critical length, $(\pi/\sqrt{2})L_a$, the current reduces to zero at high voltage as a result of the complete depletion of the i-zone. For a larger length, the current saturates at a level lower than that for infinitive length, in dependence on the length of the device.

White's model also predicts a photoelectric gain larger than unity in totally suppressed devices as the result of the additional generation contributed by the new carriers that are induced by external sources.

The three-layer devices with doped central zones, such as n^+-p-p^+ and n^+-n-p^+, cannot be analyzed with formulas but can be analyzed with computer simulations. White developed a generalized computer model for the structures that include properties of idealized models.[20] These properties include doping of the lightly doped zone, grading at contacts, Shockley-Read generation, and other factors. Calculations have been performed in one dimension for 11-μm cutoff $Hg_{1-x}Cd_x$Te homojunction diodes operating at 253 K. Doping of 8×10^{17} cm^{-3} and 5×10^{18} cm^{-3} has been assumed for the n^+- and p^+-regions. This results in 18 A/cm^2 leakage due to generation in the n^+-zone doping grading, and to the 10 A/cm^2 leakage that results from generation at the surface of the 0.6-μm-thick p^+-region.

Three principal types of device structures have been identified according to their doping level in the lightly doped central zone: (1) near intrinsic devices (π or ν), (2) devices that are already extrinsic at zero bias (n or p), and (3) an intermediate case when the doping is just extrinsic (n^-, p^-). These three cases are described below:

- *Near intrinsic devices.* These devices require a very short central zone for effective Auger suppression. This is a result of a very short ambipolar diffusion length $[L_a = (2D_p\tau_i)^{1/2} \approx 1.2\ \mu m]$.
- *Just extrinsic devices.* Even shorter zones are necessary for n^- type devices, which makes them useless for Auger-suppressed diodes. In contrast, Auger suppression is most pronounced in just extrinsic p-type devices with ambipolar diffusion length of several tens of micrometers (see Figs. 6.9 and 6.10).
- *Extrinsic devices.* The n-type devices are the least interesting for Auger-suppressed devices. The moderately doped extrinsic p-type devices show the lowest currents at low bias as the result of low initial electron concentrations. They also exhibit some Auger suppression at higher bias.

Generally, all p-type devices approach the limiting current that was generated at the contacts, which may be expected to be lower for improved contacts based on wider-gap materials. The other important result is the observation of hardness of the p-p^+ junction. This junction is significantly harder than the n^+-p^- junction.

The reduction of dark current in three-layer homojunction structures is relatively modest due to parasitic generation at interfaces and contacts. Better de-

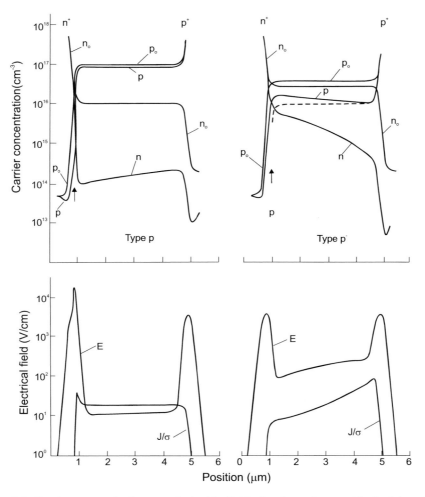

Figure 6.9 Carrier concentrations and electric fields for 4-μm long extrinsic (p) and just extrinsic (p⁻) devices. A reduction of the majority carrier concentration to the extrinsic level can be seen. (Reprinted from Ref. 20 with permission from Elsevier.)

vices can be obtained using three-layer N^+-p-P^+ (n^+-p-P^+, N-p-P, n^+-π-P^+, N^+-π-P^+).[18] Further refinements include more complex four- and five-layer structures (N^+-N-p-P-P^+ and derivative devices). These structures are capable of even better suppression of the dark current at interfaces.[2, 21–23]

Improved numerical calculations devised with software developed at some laboratories or commercially available packages make it possible to analyze complex heterostructures with bandgaps and doping-level gradings, taking into account the fundamental and nonfundamental sources of dark current.

Calculations of saturation currents as a function of residual Shockley-Read traps for three-layer structures have been reported that take into account such features as degeneracy, compensation, partial occupation of states, large deviation

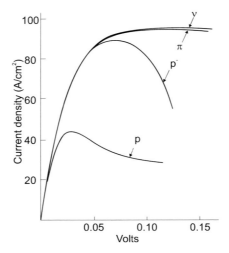

Figure 6.10 Current-voltage characteristics for 4-μm-long devices. (Reprinted from Ref. 20 with permission from Elsevier.)

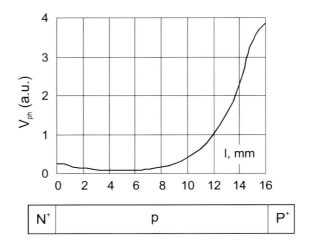

Figure 6.11 Photovoltage due to 10.6-μm-radiation scan across N^+-p-P^+ heterostructure.

from equilibrium, generation-recombination mechanisms, and graded interfaces.[24] These calculations determined the upper limit to trap concentration that can be tolerated for effective Auger suppression.

Numerical calculations have become an invaluable tool to analyze and optimize more complex heterostructures used for the equilibrium and nonequilibrium devices, including 2D and 3D structures. They also can be used to calculate the distribution of the photoelectric gain, volume density of generation-recombination noise, and photoresponse.[25]

Figure 6.11 shows the photoresponse distribution of an N^+-p-P^+ heterostructure with a 16-μm-long p-type absorber optimized for uncooled detection of 10.6 μm.[25] The photoresponse can be observed in the vicinity of the P^+-p and anisotype N^+-p heterojunctions at distances comparable to the ambipolar diffusion length. The striking difference in the response can be observed at the two junctions. Despite significant doping of the p-region, the photoresponse at the isotype P^+-p heterojunction is much larger than the anisotype N^+-p heterojunction. This confirms the relative hardness of the P^+-p junction.

6.1.4 Quantum efficiency issues of high-temperature photodiodes

In typical performance calculations, we assumed the absorption depth $1/\alpha$ was shorter than the diffusion length. This is true for optimized MWIR, but in long-wavelength devices operating at near room temperature, the diffusion length may be shorter than the absorption length. Consider an example of an uncooled 10.6-μm photodiode with an $x = 0.165$ absorber doped to approximately $N_a = 1 \times 10^{17}$ cm^{-3}. Calculations show that the ambipolar diffusion length is less than 2 μm and the absorption depth is ≈ 13 μm. This reduces the quantum efficiency to $\approx 15\%$.

There are some ways to improve this and achieve near 100% efficiency even for a diffusion length shorter than the absorption depth (see, e.g., Refs. 6, 16, 17, and 26).

Double pass of radiation with a suitable retroreflector

This simple solution may almost double the quantum efficiency for weakly absorbed radiation. For example, this solution makes it possible to achieve $\approx 25\%$ quantum efficiency in uncooled 10.6-μm photodiodes.

Optical resonant cavity

A properly designed resonant cavity makes it possible to achieve full absorption within a $\lambda/4n$ depth (≈ 0.8 μm). This depth is smaller than the diffusion length, so near-unity quantum efficiency becomes feasible. This device makes it possible to shorten the time required for photogenerated carriers to reach the space charge regions, thus improving the high-frequency response. The disadvantages are a narrow spectral response and a limited field of view.[27]

Stacked multiple-heterojunction devices[13, 15, 28–32]

A stacked multiple-heterojunction device, shown in Fig. 6.12, is based on multiple N^+-p-P^+ heterostructures. The absorption of the long-wavelength radiation occurs in the p-type absorber region of the junctions. For the best performance, the total thickness of the structure should be close to the optimum thickness $1.26/\alpha$ (or $0.63/\alpha$ for double pass). The thickness of the absorber region of each element should be less than the diffusion length. In principle, these devices make it possible to achieve near-unity quantum efficiency.

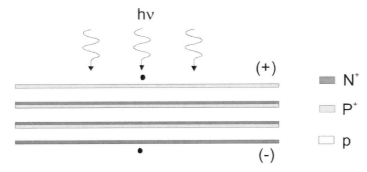

Figure 6.12 Stacked multiple-heterojunction device.

Photodiode with junction planes perpendicular to the surface of the device

A photodiode with junction planes perpendicular to the surface of the device, shown in Fig. 6.13 imposes these necessary conditions for achieving high quantum efficiency:

- For efficient absorption, the thickness (t) of the p region should be comparable to the absorption length, and the conditions should be $1.26/\alpha$ or $0.63/\alpha$ for single and double pass, respectively;
- For efficient collection, the length (l) of the p region should be shorter than the diffusion length.

The length of this device is limited to the diffusion length. The absorption depth is longer than diffusion length $l \ll t$, and the device will exhibit good absorption only for normal incidence unless the side surfaces are covered with a reflective coating.

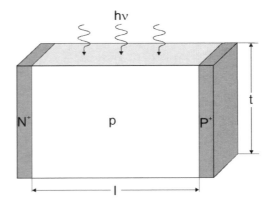

Figure 6.13 Geometry of an N^+-p-P^+ photovoltaic detector with junctions perpendicular to the front surface of the device.

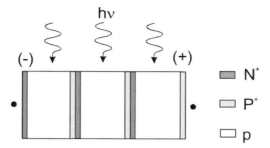

Figure 6.14 Geometry of multiple N^+-π-P^+ photovoltaic detector with junctions perpendicular to the front surface of the device.

Multiple-heterojunction device with junction perpendicular to the front surface of the device

Unlimited active area and good absorption for any incidence angle can be achieved in a multiple-heterojunction device with the junction perpendicular to the front surface of the device, as shown in Fig. 6.14.

Reducing the distance within the heavily doped contact region below the diffusion length is crucial for high-frequency operation. This also reduces the time required for the photogenerated carriers to diffuse to the wide-bandgap regions that is proportional to the square of the distance from the generation point to the contacting region. By using such a multi-junction structure comprised of cells with narrow active regions, it is possible to obtain a response time much shorter than the bulk carrier lifetime.

It is worth noting that the length (thickness) of the wider-bandgap contacting regions should be kept short compared to the length of the absorbing regions since the wider-bandgap contacting regions are optically inactive. For very short individual cells, the length of the absorbing region may become comparable to the length of the contacting region. Under these circumstances, a large proportion of the total device area may be optically inactive, which results in a reduction of the overall quantum efficiency and resulting detectivity.

6.1.5 Frequency response of $Hg_{1-x}Cd_xTe$ photodiodes

Consider the limitation to the frequency response of the three-layer heterostructure photodiodes. The frequency response is limited by two processes: (1) transport through the lightly doped absorber region, and (2) equivalent RC time constant.

Transport through the absorber region is a combination of diffusion and drift. The diffusion limited transit time is

$$\tau_{dif} = \frac{t^2}{2.4D_a}. \tag{6.26}$$

The p-type $Hg_{1-x}Cd_xTe$ is the material of choice for an absorber of a fast photodiode due to the large diffusion coefficient of electrons. The diffusion tran-

sit times are ≈ 100 ps for extrinsic p-type $Hg_{0.832}Cd_{0.168}Te$ at room temperature. Slower response should be expected in detectivity optimized devices with
an $N_a \approx 10^{17}$ cm^{-3} absorber because the ambipolar diffusion coefficient is less
than that of electrons. Further reduction can be achieved using a thinner absorber.
The double pass of radiation and optical resonance can be used to compensate the
loss of quantum efficiency, as discussed above.

Drift transport can further improve frequency response. Significant drift can be
expected in Auger-suppressed photodiodes where the electrical field in the depleted
zone is large. For fully depleted π absorbers, electrons are the minority carriers and
electron drift determines response time.[21] The drift time of 36 ps is expected for
10.6-μm devices operating at 260 K. The drift transit can be significantly slowed
down by ambipolar effects for a strong optical excitation, which causes a significant
increase in electron concentration.

The main limitation of response time is usually the *RC* time constant. For photodiodes connected to a transimpedance preamplifier with a low input resistance,
the equivalent time constant is determined by the junction capacitance and series
resistance of photodiode. The capacitance of 45-μm-diameter, 10.6-μm Auger-
suppressed photodiodes of 8 pF has been measured and is considerably greater
than what would be expected for depletion capacitance; this is thought to be due to
diffusion capacitance.[33] Series resistances of devices with a p-type bottom layer is
typically several hundred ohms. This leads to an *RC* time constant of 2.4 ns for a
45-μm-diameter photodiode.

The reduction of series resistance by almost two orders of magnitude is possible
using structures with heavily doped n-type material for the bottom layer with a
corresponding reduction of the *RC* time constant. A large reduction is expected in
an optically immersed photodiode with a very small active area.[17, 34]

As we can see, well-designed photodiodes can have very good frequency response and can be used for gigahertz detection of IR radiation.

6.2 Technology of $Hg_{1-x}Cd_xTe$ Photodiodes

6.2.1 Basic configurations of photodiodes

The basic structures used for LN-cooled $Hg_{1-x}Cd_xTe$ photodiodes are typically being implemented in the mesa, planar, and lateral configurations shown in
Fig. 6.15 (see review papers in Refs. 6 and 7). The front-side illuminated device
suffers from a reduced active area due to the contact metallization. The effective
backside operation requires thinning to approximately 10 μm to achieve high quantum efficiency and $R_o A$ product. The use of epilayers makes preparation of the
backside operation straightforward because no thinning is necessary. The epitaxial
photodiode in Fig. 6.15(c) can be both front-side and back-side illuminated. In the
latter case, all of the area above the junction can be metallized.

In principle, all devices can be front-side and back-side illuminated. Front-side
illumination requires only partial metallization of the active area, but backside illumination has important advantages. The active area can be completely metallized.

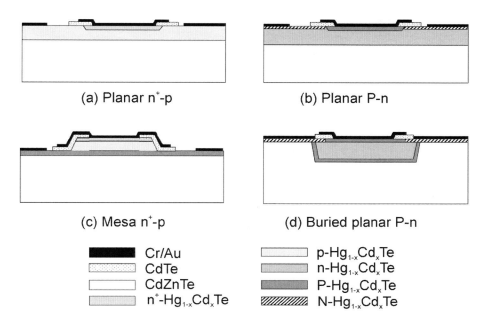

(a) Planar n⁺-p

(b) Planar P-n

(c) Mesa n⁺-p

(d) Buried planar P-n

▬ Cr/Au	▭ p-$Hg_{1-x}Cd_xTe$
▨ CdTe	▨ n-$Hg_{1-x}Cd_xTe$
▭ CdZnTe	▨ P-$Hg_{1-x}Cd_xTe$
▨ n⁺-$Hg_{1-x}Cd_xTe$	▨ N-$Hg_{1-x}Cd_xTe$

Figure 6.15 Schematic cross-sections of the most common configurations of epitaxial $Hg_{1-x}Cd_xTe$ photodiodes operating at near room temperatures.

The metallization plays the role of a reflector, and as a result, a double pass of IR radiation occurs. The substrate materials usually have lower refraction indices than $Hg_{1-x}Cd_xTe$ layers, so reflection losses are relatively low without the use of an AR coating. Next, the front-side metallization prevents unwanted emission and absorption of IR radiation. Backside-illuminated devices contacted with In bonds are typically used for large, high–fill-factor 2D arrays.

The n⁺-p structure that is so frequently used in LN-cooled devices is not so successful for uncooled operations. This structure suffers from high series resistance, lateral generation, collection of dark current, and low resistance of the n⁺-p contact.

The baseline architecture of uncooled devices is

- P⁺-p-N⁺ (or derivative) if the absorber doping $p \approx \gamma^{1/2}n_i$ is possible. This device is a typical case of uncooled and minimally cooled LWIR and MWIR devices.
- P⁺-n-N⁺ (or derivative) structures when the required absorber p-type doping or low SR generation is not feasible. This device is typical for <4 μm uncooled and cooled devices.

Series resistance of the devices is mostly due to the bottom layer resistance between the active region and the remote contact. Therefore, the use of a low resistivity N⁺-layer at the bottom of the structure is advantageous.

6.2.2 Junction formation

The p-n junctions in $Hg_{1-x}Cd_xTe$ have been formed by numerous techniques, including Hg in-and out-diffusion, impurity diffusion, ion implantation, ion milling, reactive ion etching, electron bombardment, doping during epitaxial growth, and other methods. To avoid citing hundreds of related references including those written by the author of this monograph, please see the references in these recent reviews.[6, 7]

6.2.2.1 Hg in-diffusion

Hg in-diffusion into vacancy-doped ($\approx 10^{17}$ cm^{-3}) $Hg_{1-x}Cd_xTe$, which was originally proposed by French workers, has been widely used in the past for very fast photodiodes in which a low concentration (10^{14}–10^{15} cm^{-3}) n-type region is necessary for a large depletion width and low junction capacitance. The n-type conductivity is originated from a background donor impurity. Initially, a mesa configuration was used for this type of device. Planar n^+-n-p planar structure also has been used. A 0.5-μm ZnS was sputtered onto the surface of vacancy-doped $Hg_{1-x}Cd_xTe$. After etching openings in the ZnS through a photoresist mask, and before removing the photoresist, an $\approx$10-nm In layer was sputtered onto the surface. A low-concentration n-type layer approximately 5 μm deep with a thin n^+ skin was produced by 30 min, 240°C Hg-vapor diffusion in a sealed ampoule. The photoresist lift-off technique was used to define sputtered In-Au bonding pads. A ZnS mask provided passivation around the junction perimeter. The sensitive area may be covered with AR and encapsulation coatings. A CdTe layer 0.4–0.8-μm thick also can be used as an effective barrier for Hg diffusion.

6.2.2.2 Ion implantation

The ion implantation technique has been a common method of n^+-p photodiode fabrication because it avoids heating the metallurgically sensitive material and permits precise control of the junction depth. The n^+-p structures are produced by H, Hg, Al, Be, In, and B ions' implantation into vacancy doped p-type material. The n-p junctions are located at a depth of 1–3 μm, which is large compared to the submicron range of implanted ions.

The origin of p-to-n conversion is still under discussion. The p-to-n conversion occurs for donor, acceptor, and neutral species. The integrated doping is much higher than the implanted dose. Radiation defects and Hg atoms released from the lattice due to damage induced by ion implantation are among possible causes for junction formation without post-implant annealing. Boron is possibly the most frequently used material for n^+-p photodiodes, perhaps because boron is a well-behaved shallow donor impurity in $Hg_{1-x}Cd_xTe$ and is also a standard implant for Si. The mesa diodes are made by implantation on a bare or passivated surface with subsequent etching for element separation. For planar devices, prior to implantation, the substrates are covered with dielectric layers (photoresists, ZnS, CdTe) containing openings that act as a mask for impinging ions and thus define the junction area. Implantation is typically performed at room temperature: the substrates, if oriented, are inclined to the beam axis to avoid ion channeling. Doses of 10^{12}–10^{15} cm^{-2} and energies of 30–200 keV are applied. No post-implant annealing was found to be necessary to achieve high performance, particularly at lower doses and for SWIR and MWIR devices. Some workers concluded that LWIR photodiode performance could be improved if post-implant annealing was used. The implantation effects are not annealed in vacuum at temperatures as high as 140°C.

Attempts to obtain implanted p-on-n photodiodes by implantation also proved to be successful. The junctions were formed by Au, Ag, Cu, P, and As implantation in n-type $Hg_{1-x}Cd_xTe$ followed by annealing. Annealing is usually a necessary step to activate the p-type dopants. Arsenic implantation is most widely used at present. Annealing requires 400–450°C in a high Hg pressure ambient that is necessary to activate this dopant. An example is the planar double-layer heterojunction photodiode.

6.2.2.3 Ion milling

Conversion of vacancy-doped p-type $Hg_{1-x}Cd_xTe$ to n-type during low-energy ion bombardment has become another important technique for junction fabrication. Donor ions and post-annealing are not required. The local damage caused by the impinging ions is limited to a distance of the order of the ion range. The depth of the converted zone is much larger and remains roughly proportional to the thickness of the layer removed by ion bombardment. The reason for p-to-n conversion is still under discussion, but rapid diffusion of Hg released from the lattice is commonly believed to be the cause of the junction formation. Formation of extended defects and Te antisites is also postulated.

6.2.2.4 Reactive ion etching

Reactive ion etching (RIE) is traditionally used as an anisotropic etching technique that allows for delineation of active elements in high-density structures. Exposure of p-type $Hg_{1-x}Cd_xTe$ to CH$_4$:H$_2$ RIE plasma has been shown to convert the material to n-type and has been used to fabricate n-p junctions.[35–37] The plasma-induced type of conversion may also be used with P-on-n heterostructures to insulate the junction of high-performance photodiodes.[38]

6.2.2.5 Doping during growth

Doping during epitaxial growth has become the preferred technique at present because of its inherent advantage of integrating the material growth and device processing.

High-performance photodiodes can be obtained by successive growth of doped layers using LPE from Te-rich solutions[39–43] or Hg-rich solutions,[44, 45] MBE,[46, 47] MOCVD,[48–57] and ISOVPE.[17, 58] Epitaxial techniques make it possible to grow *in situ* multilayer structures of strategic importance for uncooled photodiodes and cooled photodiodes operating at a very long wavelength. Stable and readily achievable dopants at temperatures below 300°C are required to prevent interdiffusion processes. Indium and iodine are preferred n-type dopants, and arsenic is the preferred p-type dopant for *in situ* doping as discussed in Chapter 3.

The double-layer heterojunction (DLHJ) P^+-n photodiode, which consists of an ≈10-μm base and ≈2-μm wide-gap cap layer, is probably the most common double-layer device. The advantages of the heterojunction P^+-n devices over homojunction n^+-p are especially significant for high-temperature LWIR and MWIR diodes and for cryogenically cooled photodiodes operating at a wavelength >14 μm.[51–53] Epitaxy doping is also used for more advance three-, four-, and five-layer photodiodes.[21, 59, 60]

6.2.2.6 Other techniques

In-diffusion into p-$Hg_{1-x}Cd_xTe$ has been used for n^+-p diodes.[61] Photodiodes of p-on-n structures were prepared by Au[62] and As[63] in-diffusion into n-type $Hg_{1-x}Cd_xTe$. Sputtering the doped $Hg_{1-x}Cd_xTe$ films onto $Hg_{1-x}Cd_xTe$ bulk crystals of opposite conductivity type has been used for the p-on-n and n-on-p photodiodes.[64]

6.2.3 Passivation

Passivation is a critical step in the $Hg_{1-x}Cd_xTe$ photodiode technology that greatly affects surface leakage current and device thermal stability; as a result, most manufacturers treat it as a proprietary process. Passivation of photodiodes is more difficult since the same coating must stabilize regions of n- and p-type conductivity simultaneously. The most difficult is the passivation of p-type material due to its tendency for inversion. For example, the anodic oxidation, which perfectly passivates n-type photoconductors, is not useful for passivation of p-type regions, which results in the formation of inversion layers. The reported passivation techniques involve deposition of ZnS,[23, 61, 65, 66] CdS,[67–69] fluorides,[70] SiO$_x$,[39, 71] and SiN$_x$.[72]

Recent efforts are concentrated mostly on CdTe, CdZnTe, and heterojunction passivation.[57, 66, 73–81] The layers can be sputtered, e-beam evaporated, and grown by MOCVD or MBE. Inversed ISOVPE ("negative epitaxy") also has been used.[58, 82] The CdTe passivation can be obtained during epitaxial growth,[57, 77, 80] by the post growth deposition,[78, 79] or by the anneal of the grown heterostructures in Hg/Cd vapors.[81] Directly grown *in situ* CdTe layers lead to low fixed interface charge; the indirectly grown ones are acceptable but not as good as those that are directly grown. Procedures for surface preparation prior to indirect CdTe deposition have been proposed.[76, 78]

Passivation is easier to accomplish with P-n heterojunctions where the heavily doped P layer is less vulnerable to inversion, and passivation on the n layer is relatively simple. Double-layer passivation with CdTe or wider-gap $Hg_{1-x}Cd_xTe$ overcoated with SiO$_x$, SiN$_x$, and ZnS are also frequently used.[76]

6.2.4 Contact metallization

The issues connected with contacts in photodiodes are contact resistance, contact surface recombination, contact $1/f$ noise, generation of photovoltages of the opposite sign, and long-term and thermal stability of devices. Indium has been the most important metal for n-$Hg_{1-x}Cd_x Te$ for many years,[4] while Au, Cr/Au, and Ti/Au have been used frequently for both p- and n-type material.[4] No good contacts exist for lightly doped p-type $Hg_{1-x}Cd_x Te$; all metals tend to form Schottky barriers. The problem is especially difficult for high x material. The problem could be solved by heavy doping of the semiconductor in the region close to the metallization to increase the tunnel current, but the required level of doping is difficult to achieve in practice. One practical solution is to use rapid narrowing of the bandgap at the $Hg_{1-x}Cd_x Te$-metal interface.[54, 55]

6.3 Practical Photodiodes

6.3.1 SWIR photodiodes

Interest has increased in the use of 1–3-μm devices for fiber optical communications, laser radar, rangefinders, missile seekers, laser warning devices, astronomy, earth resources, short-wavelength thermal imaging, and many other military and civilian applications.[62, 83–93] In addition to high detectivity, a wide bandwidth is usually a basic requirement. Large resistances are obtained at an operating temperature of 200–300 K, depending on the cutoff wavelength.

The present SWIR devices are typically double-layer heterojunction P-n photodiodes. The $R_o A$ product is typically determined by means other than Auger mechanisms. The dominant currents in properly prepared diodes are the diffusion bulk-generated carriers that are limited by shallow SR centers. It is believed that the SR centers are introduced during activation of As dopants rather than during MBE growth. Radiative generation also is an important source of dark current.

With a reverse bias, uncooled SWIR photodiodes exhibit avalanche breakdown. Avalanche photodiodes optimized for 1.33–2.5 μm have been reported.[4, 85, 94–96] The proximity of the bandgap values and spin-orbit splitting energies for $Hg_{1-x}Cd_x Te$ ($x = 0.6$–0.7) leads to a high hole-to-electron ionization coefficient, which enables the fabrication of low excess noise avalanche photodiodes.[94, 97, 98] High-gain 1.55-μm P-i-N photodiodes with a low excess noise factor and dark current below 10 nA have been demonstrated.[96] A 3-dB frequency well above 1 GHz has been found.

6.3.2 MWIR photodiodes

Practical MWIR photodiodes operating in temperatures of 200–300 K have been reported by several groups.[4, 86, 92, 99–101] Usually the devices are cooled with two- or three-stage TE coolers to increase the resistance of practical devices to the level

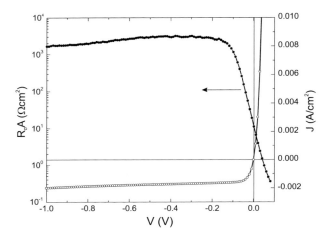

Figure 6.16 Current-voltage characteristics and differential resistance area product of a planar n^+-p photodiode optimized for operation at 4.6 μm at 230 K, where $N_a = 1 \times 10^{16}$ cm^{-3}. (Reprinted from Ref. 104 with permission from J. Antoszewski.)

where it will exceed the devices' parasitic resistances. This often depends on the device area. The operation of MWIR photodiodes can be extended to room temperature using optical immersion and other means.[102]

Figure 6.16 shows an example of I-V curves and R_oA-V dependences for n^+-p planar devices and epitaxial devices optimized for operation at 4.6 μm at 230 K. Since the diffusion length in lightly doped p-type $Hg_{1-x}Cd_xTe$ ($x \approx 0.3$) may exceed 50 μm, the perimeter region may significantly contribute to the total dark current and photocurrent. The actual size of the optically active layer can be significantly larger than the junction area for the nominal junction areas of 100×100 μm or less with gradually decreasing responsivity at the borders. This should be taken into account in R_oA product calculations. This effect is smaller in P^+-n devices as a result of the shorter diffusion length in n-type material and in mesa structures.[103] The perimeter generation of dark current can be practically eliminated in 3D structures as a result of the physical limitation of the size of the absorber region.[27]

The fundamental bulk-generated diffusion dark current dominates at low reverse, zero, and direct bias. Tunnel or avalanche multiplication currents are seen at high reverse bias, resulting in a maximum of R_oA at reverse voltages of 0.3–0.6 V, in dependence on bandgap, doping, and temperature. The zero-bias resistance is relatively low, which may be caused by the softness of the n^+-p junction at high temperatures when p-type doping is comparable with intrinsic concentration. Auger generation at n^+-p interface and Shockley-Read generation are also possible dark current sources.

Figure 6.17 compares temperature dependences of R_oA for epitaxial P^+-n (P^+-n-N^+, P^+-n-n^+) and n^+-p photodiodes. Calculated dependences are also shown for P^+-n junctions for various doping levels. Typically P^+-n devices exhibit junction resistances of approximately one order of magnitude larger than that of n^+-p homojunctions. The R_oA values for the P^+-n 4.6-μm devices at $T \approx 200$ K are

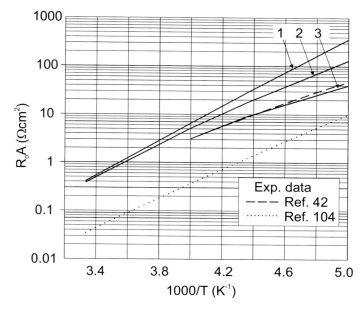

Figure 6.17 Current-voltage characteristics and normalized resistance of MWIR photodiodes optimized for operation at 4.6 μm at 200 K ($x = 0.286$). Experimental data for P-n diode taken from Ref. 42 and for n^+-p from Ref. 104. Curves 1, 2, and 3 are calculated values for a P-n device with donor concentrations of 1×10^{14} cm^{-3}, 1×10^{15} cm^{-3}, and 3×10^{15} cm^{-3}, respectively.

scattered in the range of a few tens of $\Omega\,cm^2$, reaching the limit due to Auger generation.[22, 42] Therefore, improvements at temperatures ≈ 200 K are possible by reducing absorber doping.

Series resistance is an important issue for uncooled devices with a cutoff wavelength larger than 4 μm. The main source of the series resistance is the resistance of the absorber region between the device perimeter and the remote contact. Series resistance is especially problematic of n^+-p and related structures with a p-type base where the sheet resistance of the p-layer is large (hundreds ohms) while the junction resistance is low. Even 50×50 μm uncooled 4.5-μm devices exhibit resistance dominated by the series resistance.

The series resistances are much lower in P^+-n, and especially in P^+-n-N^+ and P^+-p-N^+ devices, due to a low sheet resistivity of the N^+ layer (few ohms/square or less) so the junction resistance dominates for areas less that 1×1 mm^2. The uncooled, well-designed 4.5-μm P-n devices have resistances of a few tenths of $\Omega\,cm^2$.[102]

The internal radiative generation is an important mechanism for determining dark current in high-quality MWIR devices operating at near room temperatures.[105] This happens when a photodiode is illuminated by radiation emitted by inactive regions of a single device or other devices in an array as shown in Fig. 6.18. No radiative generation occurs with a single-element device that is shielded from outside radiation with reflective coatings.[106]

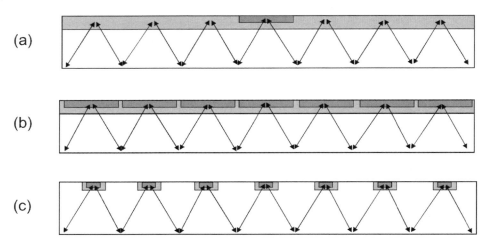

Figure 6.18 (a) Random photon exchange between active element and nonactive regions; (b) various elements of an array with element size close to the array pith; and (c) various elements with a smaller size than the array pitch.

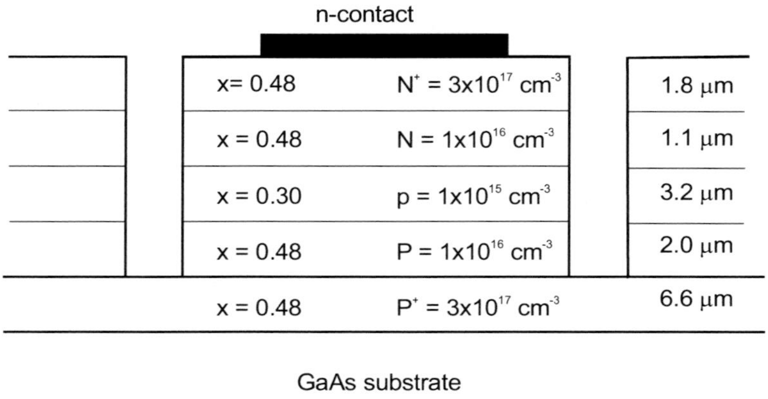

n-contact

x= 0.48	$N^+ = 3 \times 10^{17}$ cm^{-3}	1.8 µm
x = 0.48	$N = 1 \times 10^{16}$ cm^{-3}	1.1 µm
x = 0.30	$p = 1 \times 10^{15}$ cm^{-3}	3.2 µm
x = 0.48	$P = 1 \times 10^{16}$ cm^{-3}	2.0 µm
x = 0.48	$P^+ = 3 \times 10^{17}$ cm^{-3}	6.6 µm

GaAs substrate

Figure 6.19 Schematic cross-section of a five-layer heterojunction photodiode. (Reprinted from Ref. 105 with permission from TMS.)

An example is five-layer N^+-N-π-P-P^+ Hg$_{1-x}$Cd$_x$Te photodiodes (see Fig. 6.19) with adjusted doping and composition profiles; the dark currents due to Auger and contact diffusion mechanisms can be controlled, which leads to devices that are close to being radiatively limited at 240 K in the 3–5 µm spectral range.[105] The J-V curve shows good saturation of dark current (see Fig. 6.20). In Fig. 6.21, the reverse saturation current measured at a bias of 0.12 V is plotted against temperature, and this is compared with the prediction for the radiative current. Weak Auger suppression can be seen at 300 K. At low temperatures, dark current is mostly due to Auger generation in the absorber, and at interfaces, Shockley-Read generation may also play a role. The radiatively generated current increases from approximately 10% at 200 K to 53% at 240 K and

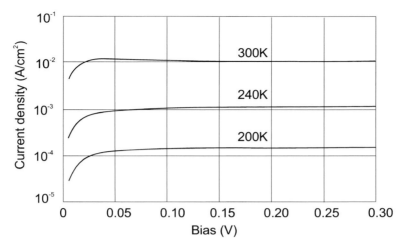

Figure 6.20 Measured dark currents for a five-layer heterojunction photodiode with 4.2-μm cutoff wavelength. (Reprinted from Ref. 105 with permission from TMS.)

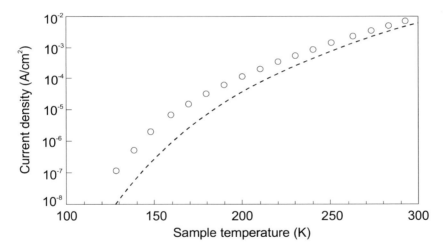

Figure 6.21 Reverse current density of 0.12 V bias as a function of temperature. (Reprinted from Ref. 105 with permission from TMS.)

65% at 300 K. The "dark" current of a detector in an array was found to depend on thermal radiation of adjacent devices; reduction of the current was observed when the neighboring devices were reverse-biased to reduce their thermal emission.

As these results show, the radiative generation may be an obstacle to achieving background-limited performance. Negative luminescence,[105] metal,[34] and dielectric[106] shields are proposed to reduce radiatively generated current.

The use of optical concentrators, such as immersion lenses, significantly improves the performance of uncooled photodiodes.[17] In addition to the main benefit of reduced physical size of the absorption region, advantages include:

- Improved ratio of the junction to the series resistance; although the junction resistance is at first approximation proportional to the area, the series resistance is less dependent on area;
- Reduced capacitance by a factor of n^2 or n^4 for hemispherical and hyperhemispherical immersion, which results in significant increases of cutoff frequency for RC-limited devices;
- Reduced bias current and heat dissipation; and
- Reduced radiative generation [see Fig. 6.19(c)] due to reduced volumes that absorb or emit radiation.

As a result, near background-limited performance is possible with uncooled and thermoelectrically cooled MWIR photodiodes.[102]

6.3.3 LWIR photodiodes

The implementation of uncooled LWIR operation presents a significant challenge due to a low junction resistance, short diffusion length, and weak radiation absorption.

Consider uncooled $\approx$10.6-μm photodiodes. The low R_oA product means that even small-sized (50×50 μm^2) and well-designed detectors exhibit less than 1 Ω zero bias junction resistances, which are well below the series resistance of a N$^+$-p-P$^+$ or P$^+$-p-N$^+$ diode.[17] While such a junction will also exhibit very low noise voltages, this does not represent a viable solution for near room temperature operation because the external resistance of the device would be dominated by parasitic resistances; consequently, the system noise would be determined by the combined Johnson noise of the parasitic resistance and the noise of the preamplifier. As a result, the uncooled photodiodes of conventional design were inferior to photoconductors and PEM detectors operated with the same conditions.

Reducing the size of the detector to 7×7 μm increases the junction resistance to ≈ 50 ohms so the junction noise approaches the noise of the series resistance. However, small devices are too small for most practical applications. The use of optical immersion can significantly increase the optical size of the detector (to 50×50 μm for CdTe hyperhemispherical lenses).

Significant increase in resistance can be achieved with thermoelectric cooling. Resistances of 11 Ω and 550 Ω are expected for 50×50 μm^2 nonimmersed and optically immersed devices at 220 K. When the ambipolar diffusion length is increased to 1.8 μm, the quantum efficiency increases to 56% with double pass of the radiation compared to 29% at 300 K. Operation at 200 K will increase the resistances to 28 Ω and 1.5 kΩ, respectively, and quantum efficiency to 60%.

Practical optically immersed LWIR photodetectors with CdZnTe or GaAs lenses have been developed and manufactured at Vigo System S.A. since 1989. The uncooled 10.6-μm devices were initially planar n$^+$-p Hg$_{1-x}$Cd$_x$Te photodiodes grown with ISOVPE. The device was supplied with a backside mirror. The immersion lens was prepared directly in CdZnTe substrates. Later, the n$^+$-p devices were replaced with advanced heterostructures having a 3D gap and doping

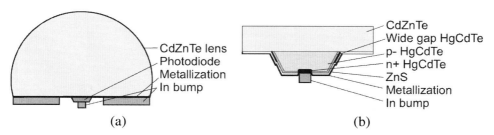

Figure 6.22 (a) Schematic cross-section of small-area, 10.6-μm photodiodes for operation at 200–300 K. (b) Enlarged view of active element.

level distribution of the mesa or buried absorber configuration (see Fig. 6.22).[34, 58] In these devices, the narrow-gap absorber is entirely surrounded by the wider-gap material, and the side walls are completely metallized.

The best devices of this type exhibit voltage responsivities approaching 100 V/W for the optically immersed devices with the apparent optical size at 50×50 μm. Despite such a small active area, the parasitic impedance of the p-type layer and contact resistance of metallization to the p-region still contribute significantly to the resulting resistance of the device (≈ 50 Ω). The Johnson-Nyquist noise-limited detectivity of the device is approximately 5×10^8 cm Hz$^{1/2}$/W. To make this detectivity practical, a preamplifier with noise voltage well below 0.5 nV/Hz$^{1/2}$ is required. Preamplifiers with this noise level can be prepared using low-noise bipolar transistors.[102] However, 2D readout integrated circuits (ROICs) with this noise level that can be used in hybrid arrays are not available at present. These detectors, which do not exhibit $1/f$ noise, can be operated at a wide frequency band starting from DC to approximately 1 GHz.

The problems with series resistances are less severe with devices operating at the short-wavelength portion of the LWIR range—where 8-μm optimized devices have resistances one order of magnitude higher. Detectivities exceeding 1×10^9 cm Hz$^{1/2}$/W and $\approx 6 \times 10^9$ cm Hz$^{1/2}$/W have been measured with uncooled $\lambda = 8.5$ μm nonimmersed and optically immersed devices.

Reflection at the CdZnTe-$Hg_{1-x}Cd_xTe$ interface to form an optical resonant cavity has been used at Vigo System S.A. in some uncooled and TE-cooled 10.6-μm devices designed for fast response times. Therefore, absorber thicknesses of ≈ 2.2 μm (less than the absorption depths) were used to reduce the diffusion time. The enhancement of responsivity was relatively modest (20–56%) as a result of a low ($\approx 4\%$) reflection coefficient at the interface. Much larger gains ($\approx 300\%$) have been obtained at the long-wavelength tail of the response when single-pass absorption is very low.

Better resonant cavities would require structures with suitable Bragg reflectors at the $Hg_{1-x}Cd_xTe$-CdZnTe interface.[27] Reducing the absorber thickness to the smallest thickness ($\lambda/4n \approx 0.75$ μm) for the optical resonant cavity would significantly reduce the dark current and increase resistance. This should lead to uncooled zero-bias PV device detectivities of $\approx 10^9$ cm Hz$^{1/2}$/W at ≈ 10 μm with very short response times. Such devices could be used for future image converters and fast receivers for optical communications.

The use of two-stage TE coolers results in increased resistance by more than one order of magnitude. Detectivities of 4×10^9 cm Hz$^{1/2}$/W have been measured at 192 K and wavelengths of 9 μm with $Hg_{1-x}Cd_x$Te photodiodes immersed in Si lenses.[107] An optically immersed 10.6-μm P$^+$-p-N$^+$ device cooled with a two-stage Peltier cooler and detectivities of $\approx 1.5 \times 10^9$ cm Hz$^{1/2}$/W has been reported.[34]

The influence of series resistance is especially pronounced in large-area devices. An unlimited increase of the detector resistance for a given active device may be attainable through the use of multiple small-volume heterostructures with junctions parallel or perpendicular to the incident radiation and connected in a series. The R_o of such a device is increased by a factor of N^2, where N is the number of heterojunctions connected in a series, over that of a single detector with the same overall volume of narrow-bandgap material. This is a consequence of the reduced thermal generation volume of each individual cell, resulting in one factor of N increment while a further factor of N arises from the series connection of the individual junction elements.

Consider a LWIR $Hg_{1-x}Cd_x$Te detector with an optical area of 1×1 mm^2 and a signal wavelength of 10.6 μm that consists of 300 P$^+$-p-n$^+$ cells, each with an optical area of 3000 μm^2. The resulting R_o is calculated to be ~ 7 kΩ and 28 kΩ at 300 K and 220 K, respectively. Correspondingly higher R_o values would be expected for devices optimized for shorter signal wavelengths and/or lower temperature operation. The end result is that there is a corresponding increase in voltage responsivity due to the increased resistance of each individual detector element and to the summation of photovoltages generated in individual cells.

Device fabrication technology, rather than fundamental physics, limits the size of each individual detector cell and the cell density. Both the dynamic resistance and voltage responsivity of the multi-junction PV detector continue to increase with increased cell density even when the length of the absorbing region within each cell is shorter than a diffusion length.

Practical large-area multiple heterojunction devices have been fabricated on epilayers grown by isothermal vapor phase epitaxy of HgCdTe and in situ As p-type doping.[28, 29, 34] More advanced multiple heterojunction devices are fabricated with MOCVD growth.[32, 54, 55] The detector structures are formed using a combination of conventional dry etching, angled ion milling, and angled thermal evaporation for contact metal deposition (see Figs. 6.23 and 6.24).

Figure 6.25 shows the performance of commercially available large-area LWIR photodetectors with multiple heterojunctions and optical immersion operating at near room temperatures. The multi-junction N$^+$-p-P $Hg_{1-x}Cd_x$Te heterostructure devices exhibit performances comparable to, or better than, existing near room temperature LWIR photoconductors and PEM detectors operated under the same conditions. Detectivities of optically immersed multiple heterostructure PV detectors exceeding 1×10^9 cm Hz$^{1/2}$/W can be achieved at a signal wavelength of 10.6 μm and for operation at 230 K.

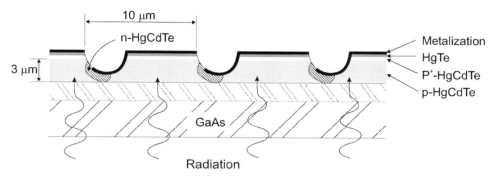

Figure 6.23 Schematic cross-section of a multiple heterojunction photovoltaic detector.

Figure 6.24 SEM image of a multi-heterojunction photodetector, where A is the mesa structure, B is a trench, C is the nonmetallized wall, and D is the nonmetallized region of the device.

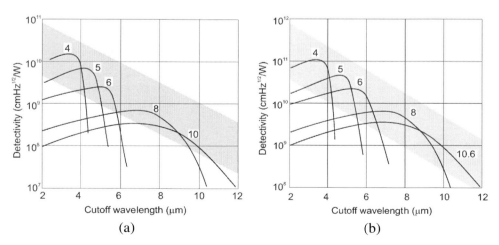

Figure 6.25 Performance of photovoltaic detectors with multiple heterojunctions and optical immersion operating at near room temperatures manufactured at Vigo System S.A.: (a) uncooled, (b) cooled with two-stage Peltier coolers. Curves show typical spectral curves while the shaded region shows the minimum and maximum performance at a given wavelength.

6.3.4 Practical stacked multiple-cell devices

Stacked multiple-cell devices require the consecutive deposition of multiple very thin N^+-p-P^+ layers. Therefore, the structures can be grown only with a low-temperature epitaxial technique that makes it possible to grow fully doped heterostructures. MBE could be an ideal technique for this purpose if *in situ* As activation is developed.

Some attempts have been made to practically implement MOCVD growth.[31, 108] Several multiple-cell LWIR photodiodes were grown. One of these was a 10.6-μm uncooled three-cell photodiode based on N^+-$Hg_{0.7}Cd_{0.3}Te$/p-$Hg_{0.83}Cd_{0.17}Te$/P^+-$Hg_{0.7}Cd_{0.3}Te$ grown at 350°C and designed for uncooled operation (see Fig. 6.26 and Table 6.2). The N^+- and P^+-layers were doped to the maximum level achievable with the applied technique. Improvement of voltage responsivity by a factor of approximately 2 has been observed, but the devices have shown relatively large resistance caused by the N^+-P^+ junctions. The problem was partially solved by narrowing the bandgap at the N^+-P^+ interface and depositing a 0.05-μm HgTe layer on top of each P^+-layer. Further efforts will be necessary to achieve the expected gains in practice.

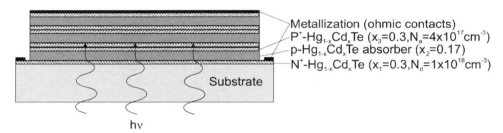

Figure 6.26 Three-cell stacked multiple detector. The back-side illuminated device is supplied with a reflector for double pass of radiation.

Table 6.2 Configuration of three-cell stacked photodiodes.

Layer	Composition x	Doping (cm^{-3})	Thickness
P^+	0.3	$\approx 4 \times 10^{17}$	0.3 μm
p	0.17	$\approx 1 \times 10^{17}$	1.8 μm
N^+	0.3	$\approx 1 \times 10^{18}$	0.2 μm
P^+	0.3	$\approx 4 \times 10^{17}$	0.2 μm
p	0.17	$\approx 1 \times 10^{17}$	1.8 μm
N^+	0.3	$\approx 1 \times 10^{18}$	0.2 μm
P^+	0.3	$\approx 4 \times 10^{17}$	0.3 μm
p	0.17	$\approx 1 \times 10^{17}$	2.0 μm
N^+	0.3	$\approx 1 \times 10^{18}$	5.0 μm
CdTe	–	–	5.0 μm
GaAs	–	–	0.4 mm

6.3.5 Auger-suppressed photodiodes

For a long time a group of British scientists was the only group to implement Auger-suppressed photodiodes. Practical realization of Auger-suppressed photodiodes was impossible because there had been a lack of technology to obtain wide-gap P contacts to a $Hg_{1-x}Cd_x$Te absorber region. Therefore, the first Auger-suppressed devices were III-V heterostructures with an InSb absorber.[18, 109–111] The first reported $Hg_{1-x}Cd_x$Te Auger extracting diodes were so-called proximity-extracting diode structures (see Fig. 6.27), in which additional guard reverse-biased n^+-n^- junctions were placed in the current path between the p^+ and n^+ regions to intercept the electron injected from the p^+ region. Philips Components Ltd. fabricated practical proximity-extracting devices in both a linear and cylindrical geometry.[18] The devices were fabricated from bulk-grown $Hg_{1-x}Cd_x$Te with a cut-off wavelength of 9.3 μm at 200 K ($x = 0.2$). The acceptor concentration of 8×10^{15} cm^{-3} was due to native doping. The n^+-regions were fabricated by ion milling. The I-V characteristics of such structures were complex and difficult to interpret because of the occurrence of bipolar transistor action and impact ionization, but the general features were predictable when the standard transistor modeling were applied. A current reduction at a factor of 48 was obtained by biasing the guard junction, but the extracted current was much greater than that predicted for an Auger-suppressed, Shockley-Read limited case. This is possibly due to a surface-generated current.

The measured 500 K blackbody detectivity of a 320 μm^2 optical area device at a modulation frequency of 20 kHz was 1×10^9 cm Hz$^{1/2}$/W, which is the best ever measured value for any photodetector operated under similar conditions.

The use of a wide-gap P contact is a straightforward way to eliminate harmful thermal generation in the p-type region. The practical realization of such devices would require well-established multilayer epitaxial technology capable of growing high-quality heterostructures with complex gap and doping profiles. This

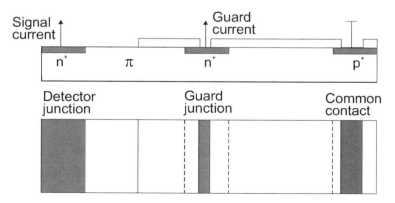

Figure 6.27 Schematic structure of a proximity-extracting photodiode. (Reprinted from Ref. 18 with permission from IOP.)

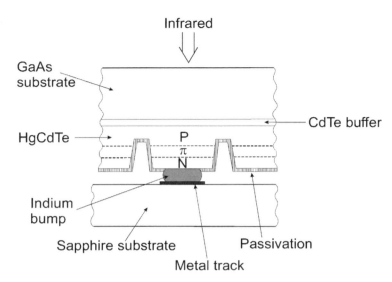

Figure 6.28 A cross-section through part of a mesa diode array. (Reprinted from Ref. 21 with permission from TMS.)

technology became available in the early 1990s. Since then, three-layer n[+]-π-P[+], N[+]-π-P[+] heterostructural photodiodes have been demonstrated[112] and gradually improved.[21, 22, 111, 113–115] The multilayer MOCVD process (IMP) has been used to grow the three-layer devices' heterostructures on CdZnTe and GaAs substrates.[22] Arsenic has been used for acceptor doping and iodine for donor doping. The doping has been introduced during CdTe cycles of IMP growth. Individual mesa diodes were defined by etching slots (Fig. 6.28) between adjacent devices.

The arrangement of the photodiode in the optical resonant cavity makes it possible to use a thin extracted zone without a loss in quantum efficiency, which is also favorable for reducing the saturation current. Additionally, this results in minimizing noise and bias power dissipation.

For low bias, the device behaves as a linear resistor until the electric field exceeds the critical value for exclusion or extraction (Fig. 6.29). Then the current drops sharply from its maximum value I_{max}, which further increases the voltage and gradually decreases the current to minimum I_{min}. At high voltages, the current increase is due to diode breakdown. As a result, the dynamic resistance increases to high values in the regions close to the transition range. The devices have shown negative resistances for temperatures above 190 K. The minimum dark current (Fig. 6.30) follows the empirical expression

$$I_{min} = 1.3 \times 10^4 \exp\left(-\frac{qE_g}{kT}\right) \quad (E_g \text{ in eV}, \lambda \text{ in micrometers}). \qquad (6.27)$$

When comparing the trend line for experimental results with expected dark current, significant Auger suppression is seen for the wavelength $\lambda > 6$ μm. The

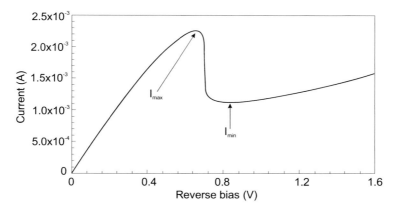

Figure 6.29 An example of current voltage characteristics of a P-p-N heterostructure showing the positions of I_{max} and I_{min}. (Reprinted from Ref. 116 with permission from TMS.)

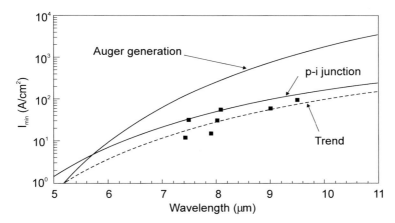

Figure 6.30 I_{min} in P-p-N diodes as a function of cutoff wavelength. The trend line is shown as a dashed line. The results are compared with the leakage current expected for a p^+-i junction and also with the current that would be generated in the π-region in the absence of Auger suppression. (Reprinted from Ref. 21 with permission from TMS.)

shot noise detectivity calculated as $R_i(2q\,J_{min})^{-1/2}$ is

$$D^* = 1.2 \times 10^7 \eta\lambda \exp\left(\frac{qE_g}{kT}\right) \quad (E_g \text{ in eV, } \lambda \text{ in micrometers}). \quad (6.28)$$

Figure 6.31 shows detectivity of Auger-suppressed N-p-P photodiodes as a function of wavelength and temperature calculated according to Eq. (6.28). This corresponds to $\lambda = 10.6$ μm detectivities of $\approx 1 \times 10^9$ cm Hz$^{1/2}$/W and $\approx 3 \times 10^9$ cm Hz$^{1/2}$/W at temperatures of 300 K and 200 K, respectively. The measured quantum efficiency often exceeded 100%, possibly as the result of the mixed conductivity effects, impact ionization, or lateral collection.

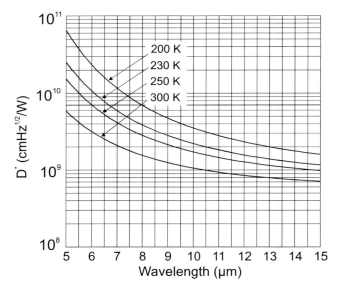

Figure 6.31 Shot noise limited detectivity of Auger-suppressed N-p-P photodiodes as a function of wavelength and temperature. Quantum efficiency of 80% has been assumed.

Auger-suppressed photodiodes have been used as 10.6-μm heterodyne receivers.[116] A noise equivalent power (NEP) of 2×10^{-19} W/Hz at frequencies above the $1/f$ knee was deduced from heterodyne experiments for detectors operating at 260 K with 0.3 mW local oscillator power.

Auger-suppressed N-π-P photodiodes grown by MBE using Ag as the acceptor and In as the donor dopant were also demonstrated.[60] The results show that fast Ag diffusion at the growth temperature does not necessarily preclude the use of Ag doping in devices. The onset of Auger suppression appears to occur at different voltages for the two different mesa diameters, but this is due to the different voltage drop across the series resistances for the two sets of diodes. The minimum reverse current density is similar to that obtained in MOVPE-grown material at the same ≈ 9 μm cutoff wavelength at 300 K. Quantum efficiencies exceeding 100% have been measured and attributed to carrier multiplication due to the relatively high bias across the sample or to mixed conduction effects.

More recently, improved N^+-N^--π-P^--P^+ $Hg_{1-x}Cd_xTe$ heterostructures with refined bandgap and doping profiles were reported.[22, 23] The N and P layers were used to minimize the generation of different composition between regions. The doping of the π-region was $\approx 2 \times 10^{15}$ cm^{-3}. The thickness of the π-region was typically 3 μm wide, which ensured good quantum efficiency with a mesa contact acting as a reflector. All the layers were annealed at 220°C for 60 hours in Hg-rich nitrogen having the same temperature as the Hg reservoir. Mesa structures passivated with CdTe and ZnS were received. Several pre-passivation clean procedures have been tried: anodizing, HBr/Br etching, and citric acid etching. Some of the CdTe passivated devices were annealed at 220°C for 30 hours in Hg-rich nitrogen for the purpose of interdiffusing CdTe/HgCdTe to produce a graded interface.

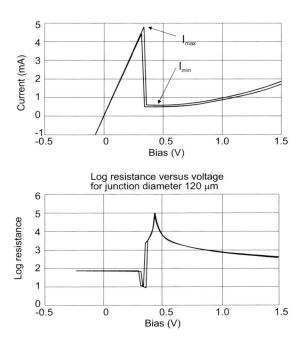

Figure 6.32 Room temperature *I-V* and differential resistance of an Auger-suppressed photodiode as a function of bias. (Reprinted from Ref. 23 with permission from TMS.)

The I-V curves with a rapid decrease of dark current from I_{max} to I_{min} have been observed (see Fig. 6.32). Some hysteresis effects due to series resistance can be seen. Peak-to-valley ratios up to 35 have been observed with extracted saturation current densities of 10 A/cm^2 and shot noise detectivities of $\approx 3 \times 10^9$ cm Hz$^{1/2}$/W in uncooled ≈ 10 µm cutoff wavelength diodes, which is an improvement by a factor of approximately 3 compared to earlier results.[23]

The residual dark current is a result of residual Auger generation (in the depleted zones and at interfaces), Shockley-Read generation in the whole structure region, and the surface leakage currents.

The present Auger-suppressed devices exhibit a high low-frequency noise with $1/f$ knee frequencies ranging from 100 to a few MHz for ≈ 10 µm devices at room temperature. This reduces their SNR at frequencies of ≈ 1 kHz to a level below that for equilibrium devices. The $1/f$ noise remains the main obstacle to achieving background-limited NETD in 2D arrays at near room temperatures.

The $1/f$ noise level is much lower in MWIR devices, and they appear to be capable of achieving useful D^* values for imaging applications in very high frame rate or high speed choppers.[3, 117] Typically, the low-frequency noise current is proportional to the bias current with $a \approx 2 \times 10^{-4}$.

Recently, the HOTEYE thermal imaging camera (based on a 320×256 FPA with 4-µm cut-off wavelength at 210 K) has been demonstrated (see Fig. 6.33).[117] The pixel structures were P$^+$-p-N$^+$ starting from the GaAs substrate. The histogram peaks were measured at an NEDT of around 60 mK with $f/2$ optics. When

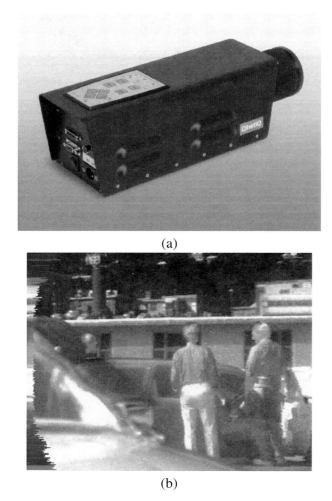

(a)

(b)

Figure 6.33 The HOTEYE camera: (a) housed in a compact encapsulation $12 \times 12 \times 30$ cm (including lens); (b) typical image. (Reprinted from Ref. 117.)

the optimum bias was selected for each pixel, an improvement in the NEDT histogram was observed. However, this procedure was problematic. A sample of the HOTEYE camera image is shown in Fig. 6.33(b).

The reason for the large $1/f$ noise is not clear.[14, 23] Traps in depletion region,[118] high field areas,[119] hot electrons,[120] and background optical generation[121] are among some of the possible reasons. Attempts to find the source of the $1/f$ noise using the perimeter-area analysis resulted in partly contradictory results.[23] It was found that for a $\approx$50-µm-diameter diode, the perimeter contributes 33% of the leakage and 57% of the noise power. Some reduction of the $1/f$ noise has been achieved with low-temperature annealing. CdTe passivation was found to give better stability than ZnS.

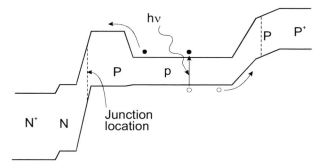

Figure 6.34 Schematic of a six-layer extracted photodiode with a wider-gap P layer located between the absorber and the N-N$^+$ contact.

It may be possible to reduce the $1/f$ noise by reducing the dark current and/or the I_n/I ratio with improved material technology and better device design.

For MWIR devices, the origin of the noise is suspected to be the depletion region, exponentially dependent on bandgap:

$$\text{noise} \propto \exp\left(-\frac{E_g}{2kT}\right).^{14}$$

A modified junction-enhanced semiconductor structure (JESS) with an additional electron barrier near the N$^+$ side of the active region (Fig. 6.34) has been proposed to reduce $1/f$ noise.[14] Excellent reverse bias characteristics with increased breakdown voltage and high quantum efficiency were observed in this structure, but thus far no significant improvement in low-frequency performance has been obtained.

At present, an exotic hetero-junction photo-JFET with a wide-bandgap gate and advanced heterostructures with separated extraction and signal contacts is under consideration as a promising HOT device for 2D arrays, but no clear pathway to eliminating the low-frequency problem has been demonstrated yet.[122]

References

1. W. D. Lawson, S. Nielsen, E. H. Putley, and A. S. Young, "Preparation and properties of HgTe-CdTe," *J. Phys. Chem. Solids* **9**, 325–329 (1959).
2. T. Elliott, N. T. Gordon, and A. M. White, "Towards background-limited, room-temperature, infrared photon detectors in the 3–13 μm wavelength range," *Appl. Phys. Lett.* **74**, 2881–2883 (1999).
3. J. Piotrowski, Z. Nowak, J. Antoszewski, C. Musca, J. Dell, and L. Faraone, "A novel multi-heterojunction HgCdTe long-wavelength infrared photovoltaic detector for operation under reduced cooling conditions," *Semicond. Sci. Technol.* **13**, 1209–1214 (1998).

4. M. B. Reine, A. K. Sood, and T. J. Tredwell, "Photovoltaic infrared detectors," *Semiconductors and Semimetals,* Vol. 18, R. K. Willardson and A. C. Beer (Eds.), Academic Press, New York, 201–311 (1981).

5. C. T. Elliott and N. T. Gordon, "Infrared detectors," in *Handbook on Semiconductors,* Vol. 4, 841–936, C. Hilsum (Ed.), North-Holland, Amsterdam (1993).

6. J. Piotrowski, "$Hg_{1-x}Cd_xTe$ infrared photodetectors," in *Infrared Photon Detectors,* SPIE Press, Bellingham, WA, 391–494 (1995).

7. A. Rogalski, K. Adamiec, and J. Rutkowski, *Narrow-Gap Semiconductor Photodiodes,* SPIE Press, Bellingham, WA (2000).

8. P. R. Bratt, "HgCdTe heterojunctions," *J. Vac. Sci. Technol. A* **1**, 1687–1693 (1983).

9. P. Migliorato and A. M. White, "Common anion heterojuctions: CdTe-CdHgTe," *Solid State Electron.* **26**, 65–72 (1983).

10. P. R. Bratt and T. N. Casselman, "Potential barriers in HgCdTe heterojunctions," *J. Vac. Sci. Technol. A* **3**, 238–245 (1985).

11. K. Zanio and K. Kay, "Modeling of HgCdTe heterojunction devices," *Mat. Res. Soc. Proc.* **90**, 39–46 (1987).

12. K. Kosai and W. A. Radford, "Influence of barriers on charge transport across HgCdTe heterojunctions," *J. Vac. Sci. Technol. A* **8**, 1254–1259 (1990).

13. J. Piotrowski and W. Gawron, "Extension of long-wavelength IR photovoltaic detector operation to near room-temperatures," *Infrared Phys. Technol.* **36**, 1045–1051 (1995).

14. M. K. Ashby, N. T. Gordon, C. T. Elliott, C. L. Jones, C. D. Maxey, L. Hipwood, and R. Catchpole, "Novel $Hg_{1-x}Cd_xTe$ device structure for higher operating temperature detectors," *J. Electron. Mater.* **32**, 667–671 (2003).

15. J. Piotrowski, "Uncooled operation of IR photodetectors," *Opto-Electr. Rev.* **12**, 11–122 (2004).

16. J. Piotrowski and A. Rogalski, "Uncooled long wavelength infrared photon detectors," *Infrared Phys. Technol.* **46**, 115–131 (2004).

17. J. Piotrowski, W. Galus, and M. Grudzień, "Near room-temperature IR photodetectors," *Infrared Phys.* **31**, 1–48 (1991).

18. C. T. Elliott, "Non-equilibrium modes of operation of narrow-gap semiconductor devices," *Semiconductor Sci. Technol.* **5**, S30–S37 (1990).

19. M. White, "Auger suppression and negative resistance in low gap diode structures," *Infrared Phys.* **26**, 317–324 (1986).

20. M. White, "Negative resistance with Auger suppression in near-intrinsic low-bandgap photodiode structure," *Infrared Phys.* **27**, 361–369 (1987).

21. C. T. Elliott, N. T. Gordon, R. S. Hall, T. J. Phillips, A. M. White, C. L. Jones, C. D. Maxey, and N. E. Metcalfe, "Recent results on MOVPE grown heterostructure devices," *J. Electron. Mater.* **25**, 1139–1145 (1996).

22. D. Maxey, C. L. Jones, N. E. Metcalfe, R. Catchpole, M. R. Houlton, A. M. White, N. T. Gordon, and C. T. Elliott. "Growth of fully doped $Hg_{1-x}Cd_xTe$ heterostructures using a novel iodine doping source to achieve

improved device performance at elevated temperatures," *J. Electron. Mater.* **25**, 1276–1285 (1996).

23. C. L. Jones, N. E. Metcalfe, A. Best, R. Catchpole, C. D. Maxey, N. T. Gordon, R. S. Hall, T. Colin, and T. Skauli, "Effect of device processing on 1/f noise in uncooled, Auger-suppressed CdHgTe diodes," *J. Electron. Mater.* **27**, 733–739 (1998).

24. P. Davis, C. T. Elliott, and A. M. White, "Current gain in photodiode structures," *Infrared Phys.* **31**, 575–577 (1991).

25. J. Piotrowski, A. Jóźwikowska, K. Jóźwikowski, and R. Ciupa. "Numerical analysis of long wavelength extracted photodiodes," *Infrared Phys.* **34**, 565–572 (1993).

26. J. M. A. Kinch, "Fundamental physics of infrared detector materials," *J. Electron. Mater.* **29**, 809–817 (2000).

27. M. Sioma and J. Piotrowski, "Modeling and optimization of high temperature detectors of long wavelength IR radiation with optical resonant cavity," *Opto-Electr. Rev.* **12**, 157–160 (2004).

28. C. Musca, J. Antoszewski, J. Dell, L. Faraone J. Piotrowski, and Z. Nowak, "Multi-junction HgCdTe long wavelength infrared photovoltaic detector for operation at near room temperature," *J. Electron. Mater.* **27**, 740–746 (1998).

29. J. Piotrowski, Z. Nowak, J. Antoszewski, C. Musca, J. Dell, and L. Faraone, "A novel multi-heterojunction HgCdTe long-wavelength infrared photovoltaic detector for operation under reduced cooling conditions," *Semicond. Sci. Technol.* **13**, 1209–1214 (1998).

30. W. Gawron and A. Rogalski, "HgCdTe buried multi-junction photodiodes fabricated by the liquid phase epitaxy," *Infrared Phys. & Technol.* **43**, 157–163 (2002).

31. J. Piotrowski, P. Brzozowski, and K. Jóźwikowski, "Stacked multijunction photodetectors of long wavelength radiation," *J. Electron. Mater.* **32**, 672–676 (2003).

32. A. Piotrowski, P. Madejczyk, W. Gawron, K. Kłos, J. Pawluczyk, M. Grudzień, J. Piotrowski, and A. Rogalski, "Growth of MOCVD HgCdTe heterostructures for uncooled infrared photodetectors," *Bull. Pol. Ac.: Tech.* **53**, 139–149 (2005).

33. T. J. Phillips and N. T. Gordon, "Negative diffusion capacitance in Auger-suppressed HgCdTe heterostructure diodes," *J. Electron. Mater.* **25**, 1151–1156 (1995).

34. J. Piotrowski, M. Grudzień, Z. Nowak, Z. Orman, J. Pawluczyk, M. Romanis, and W. Gawron, "Uncooled photovoltaic Hg$_{1-x}$Cd$_x$Te LWIR detectors," *Proc. SPIE* **4130**, 175–184 (2000).

35. O. P. Agnihorti, H. C. Lee, and K. Yang, "Plasma induced type conversion in mercury cadmium telluride," *Semicond. Sci. Technol.* **17**, R11–R19 (2002).

36. J. M. Dell, J. Antoszewski, M. H. Rais, C. Musca, J. K. White, B. D. Nener, and L. Faraone, "HgCdTe mid-wavelength IR photovoltaic detectors fabricated using plasma induced junction technology," *J. Electron. Mater.* **29**, 841–848 (2000).

37. J. K. White, J. Antoszewski, P. Ravinder, C. A. Musca, J. M. Dell, L. Faraone, and J. Piotrowski, "Passivation effects on reactive ion etch formed n-on-p junctions in HgCdTe," *J. Electron. Mater.* **31**, 743–748 (2002).

38. T. Nguyen, C. A. Musca, J. M. Dell, R. H. Sewell, J. Antoszewski, J. K. White, and L. Faraone, "HgCdTe long-wavelength IR photovoltaic detectors fabricated using plasma induced junction formation technology," *J. Electron. Mater.* **32**, 615–621 (2003).

39. M. Lanir and K. J. Riley, "Performance of PV HgCdTe arrays for 1–14-µm applications," *IEEE Trans. Electr. Dev.* **ED-29**, 274–279 (1982).

40. C. C. Wang, "Mercury cadmium telluride junctions grown by liquid phase epitaxy," *J. Vac. Sci. Technol.* **B9**, 1740–1745 (1991).

41. L. J. Kozlowski, S. L. Johnston, W. V. McLevige, A. H. B. Vandervyck, D. E. Cooper, S. A. Cabelli, E. R. Blazejewski, K. Vural, and W. E. Tennant, "128 × 128 PACE 1 HgCdTe hybrid FPAs for thermoelectrically cooled applications," *Proc. SPIE* **1685**, 193–203 (1992).

42. A. I. D'Souza, J. Bajaj, R. E. DeWames, D. D. Edwall, P. S. Wijewarnasuriya, and N. Nayar, "MWIR DLPH HgCdTe photodiode performance dependence on substrate material," *J. Electron. Mater.* **27**, 727–732 (1998).

43. A. I. D'Souza, M. G. Stapelbroek, E. R. Bryan, A. L. Vinson, J. D. Beck, M. A. Kinch, Ch.-F. Wan, and J. E. Robinson, "Composite substrate for large-format HgCdTe IRFPA," *Proc. SPIE* **5074**, 146–153 (2003).

44. T. Tung, L. V. DeArmond, R. F. Herald, P. E. Herning, M. H. Kalisher, D. A. Olson, R. F. Risser, A. P. Stevens, and S. J. Tighe, "State of the art of Hg-melt LPE HgCdTe at Santa Barbara Research Center," *Proc. SPIE* **1735**, 109–131 (1992).

45. G. N. Pultz, P. W. Norton, E. E. Krueger, and M. B. Reine, "Growth and characterization of p-on-n HgCdTe liquid-phase epitaxy heterojunction material for 11–18 µm applications," *J. Vac. Sci. Technol.* **B9**, 1724–1730 (1991).

46. J. M. Arias, S. H. Shin, J. G. Pasko, S. H. Shin, L. O. Bubulac, R. E. DeWames, and W. E. Tennant, "Molecular beam epitaxy growth and *in situ* arsenic doping of p-on-n HgCdTe heterojunctions," *J. Appl. Phys.* **69**, 2143–2148 (1991).

47. J. M. Arias, J. G. Pasko, M. Zandian, S. H. Shin, G. M. Williams, L. O. Bubulac, R. E. DeWames, and W. E. Tennant, "MBE HgCdTe heterostructure p-on-n planar infrared photodiodes," *J. Electron. Mater.* **22**, 1049–1053 (1993).

48. C. F. Byrne and P. Knowles, "Infrared photodiodes formed in mercury cadmium telluride grown by MOCVD," *Semicond. Sci. Technol.* **3**, 377–381 (1988).

49. K. Ghandhi, K. K. Parat, H. Ehsani, and I. B. Bhat, "High quality planar HgCdTe photodiodes fabricated by the organometallic epitaxy (Direct Alloy Growth Process)," *Appl. Phys. Lett.* **58**, 828–830 (1991).

50. P. Mitra, T. R. Schimert, F. C. Case, and Y. L. Tyan, "Iodine doping and MOCVD *in-situ* growth of HgCdTe p-on-n heterojunctions," *Proc. SPIE* **2228**, 96–105 (1994).

51. P. Mitra, T. R. Schimert, F. C. Case, R. Starr, M. H. Weiler, M. Kestigian, and M. B. Reine, "Metalorganic chemical vapor deposition of HgCdTe for photodiode applications," *J. Electron. Mater.* **24**, 661–668 (1995).

52. V. Rao, H. Ehsani, I. B. Bhat, M. Kestigian, R. Starr, M. H. Weiler, and M. B. Reine, "Metalorganic vapor phase epitaxy *in-situ* growth of p-on-n and n-on-p $Hg_{1-x}Cd_xTe$ junction photodiodes using tertiarybutylarsine as the acceptor source," *J. Electron. Mater.* **24**, 437–443 (1995).

53. G. Bahir, V. Garber, and T. Dust, "Characterization of a new planar process for implementation of p-on-n HgCdTe heterostructure photodiodes," *J. Electron. Mater.* **30**, 704–710 (2001).

54. A. Piotrowski, P. Madejczyk, W. Gawron, K. Kłos, M. Romanis, M. Grudzień, A. Rogalski, and J. Piotrowski, "MOCVD growth of $Hg_{1-x}Cd_xTe$ heterostructures for uncooled infrared photodetectors," *Opto-Electr. Rev.* **12**, 453–458 (2004).

55. A. Piotrowski, P. Madejczyk, W. Gawron, K. Kłos, J. Pawluczyk, M. Grudzień, J. Piotrowski, and A. Rogalski, "MOCVD HgCdTe heterostructures for uncooled infrared photodetectors," *Proc. SPIE* **5732**, 273–284 (2005).

56. J. Giess, M. A. Glover, N. T. Gordon, A. Graham, M. K. Haigh, J. E. Hails, D. J. Hall, and D. J. Lees, "Dual-waveband infrared focal plane arrays using MCT grown by MOVPE on silicon substrates," *Proc. SPIE* **5783**, 316–324 (2005).

57. D. J. Hall, L. Buckle, N. T. Gordon, J. Giess, J. E. Hails, J. W. Cairns, R. M. Lawrence, A. Graham, R. S. Hall, C. Maltby, and T. Ashley, "High-performance long-wavelength HgCdTe infrared detectors grown on silicon substrates" *Appl. Phys. Lett.* **84**, 2113–2115 (2004).

58. J. Piotrowski and M. Razeghi, "Improved performance of IR photodetectors with 3D gap engineering," *Proc. SPIE* **2397**, 180–192 (1995).

59. C. T. Elliott, N. J. Gordon, R. S. Hall, T. J. Phillips, C. L. Jones, B. E. Matthews, C. D. Maxey, and N. E. Metcalfe, "Metal organic vapor phase epitaxy (MOVPE) grown heterojunction diodes in $Hg_{1-x}CdxTe$," *Proc. SPIE* **2269**, 648–657 (1994).

60. T. Skauli, H. Steen, T. Colin, P. Helgesen, S. Lovold, C. T. Elliott, N. T. Gordon, T. J. Phillips, and A. M. White, "Auger suppression in CdHgTe heterostructure diodes grown by molecular beam epitaxy using silver as acceptor dopant," *Appl. Phys. Lett.* **68**, 1235–1237 (1996).

61. P. Migliorato, R. F. C. Farrow, A. B. Dean, and G. M. Williams, "CdTe/HgCdTe indium-diffused photodiodes," *Infrared Phys.* **22**, 331–336 (1982).

62. T. Koehler and P. J. McNally, "(Hg,Cd)Te photodiodes for communication systems," *Opt. Eng.* **13**, 312–315 (1974).

63. J. Rutkowski, A. Rogalski, J. Piotrowski, and J. Pawluczyk, "Arsenic diffused p^+-n HgCdTe photodiodes," *Proc. SPIE* **1845**, 171–175 (1992).

64. A. Zozime, G. C. Solal, and F. Bailly, "Growth of thin films of $Cd_xHg_{1-x}Te$ solid solutions by cathodic sputtering in a mercury vapor plasma," *Thin Solid Films* **13**, 139–152 (1980).

65. D. L. Spears, "Planar HgCdTe quadrantal heterodyne arrays with GHz response at 10,6 μm," *Infrared Phys.* **17**, 5–8 (1977).

66. C. A. Musca, J. Antoszewski, J. M. Dell, L. Faraone, and S. Terterian, "Planar p-on-n HgCdTe heterojunction mid-wavelength infrared photodiodes formed using plasma-induced junction isolation," *J. Electron. Mater.* **32**, 622–626 (2003).

67. Y. Nemirovsky and G. Bahir, "Passivation of mercury cadmium telluride surfaces," *Vac. Sci. Technol. A* **7**, 450–459 (1989).

68. Y. Nemirovsky and L. Burstein, "Anodic sulfide films on $Hg_{1-x}Cd_xTe$," *Appl. Phys. Lett.* **44**, 443–444 (1984).

69. Y. Nemirovsky, L. Burstein, and I. Kidron, "Interface of p-type $Hg_{1-x}Cd_xTe$ passivated with native sulfides," *J. Appl. Phys.* **58**, 366–373 (1985).

70. E. Weiss and C. R. Helms, "Composition, growth mechanism, and oxidation of anodic fluoride films on $Hg_{1-x}Cd_xTe$ ($x \approx 0.2$)," *J. Electrochem. Soc.* **138**, 993–999 (1991).

71. Y. Shacham-Diamand, T. Chuh, and W. G. Oldham, "The electrical properties of Hg-sensitized "Photox"-oxide layers deposited at 80°C," *Solid-St. Electron.* **30**, 227–233 (1987).

72. N. Kajihara, G. Sudo, Y. Miyamoto, and K. Tonikawa, "Silicon nitride passivant for HgCdTe n^+-p diodes," *J. Electron. Soc.* **135**, 1252–1255 (1988).

73. Y. Nemirovsky, "Passivation with II-VI compounds," *J. Vac. Sci. Technol. A* **8**, 1185–1187 (1990).

74. G. Sarusi, G. Cinader, A. Zemel, D. Eger, and Y. Shapira, "Application of CdTe epitaxial layers for passivation of p-type $Hg_{0.77}Cd_{0.23}Te$," *J. Appl. Phys.* **71**, 5070–5075 (1992).

75. O. P. Agnihotri, C. A. Musca, and L. Faraone, "Current status and issues in the surface passivation technology of mercury cadmium telluride infrared detectors," *Semicond. Sci. Technol.* **13**, 839–845 (1998).

76. J. White, J. Antoszewski, R. Pal, C. A. Musca, J. M. Dell, L. Faraone, and J. Piotrowski, "Passivation effects on reactive-ion-formed n-on-p junctions in HgCdTe," *J. Electron. Mater.* **31**, 743–748 (2002).

77. G. Bahir, V. Ariel, V. Garber, D. Rosenfeld, and A. Sher, "Electrical properties of epitaxially grown CdTe passivation for long-wavelength HgCdTe photodiodes," *Appl. Phys. Lett.* **65**, 2725–2727 (1994).

78. Y. Nemirovsky, N. Amir, D. Goren, G. Asa, N. Mainzer, and E. Weiss, "The interface of metalorganic chemical vapor deposition-CdTe/HgCdTe," *J. Electron. Mater.* **24**, 1161–1167 (1995).

79. A. Mestechkin, D. L. Lee, B. T. Cunningham, and B. D. Mac Leod, "Bake stability of long-wavelength infrared HgCdTe photodiodes," *J. Electron. Mater.* **24**, 1183–1187 (1995).

80. Y. Nemirovsky, N. Amir, and L. Djaloshinski, "Metalorganic chemical vapor deposition CdTe passivation of HgCdTe," *J. Electron. Mater.* **25**, 647–654 (1995).

81. S. Y. An, J. S. Kim, D. W. Seo, and S. H. Suh, "Passivation of HgCdTe p-n diode junction by compositionally graded HgCdTe formed by annealing in a Cd/Hg atmosphere," *J. Electron. Mater.* **31**, 683–687 (2002).

82. K. Adamiec, M. Grudzien, Z. Nowak, J. Pawluczyk, J. Piotrowski, J. Antoszewski, J. Dell, C. Musc, and L. Faraone, "Isothermal vapor phase epitaxy as a versatile technology for infrared photodetectors," *Proc. SPIE* **2999**, 34–43 (1997).

83. A. N. Kohn and J. J. Schlickman, "1–2 micron (Hg,Cd)Te photodetectors," *IEEE Trans. Electr. Dev.* **ED-16**, 885–890 (1969).

84. D. A. Soderman and W. H. Pinkston, "(Hg,Cd)Te photodiode laser receivers for the 1–3 μm spectral region," *Appl. Opt.* **11**, 2162–2168 (1972).

85. S. H. Shin, J. G. Pasko and D. T. Cheung, "Epitaxial HgCdTe/CdTe photodiodes for 1–3 μm spectral region," *Proc. SPIE* **272**, 27–31 (1981).

86. K. Vural, "Mercury cadmium telluride short- and medium-wavelength infrared staring focal plane arrays," *Opt. Eng.* **26**, 201–208 (1987).

87. L. M. Smith, J. Thompson, P. Mackett, T. Jenkin, T. Nguyen Duy, P. Gori, A. Cetronio, C. Lanzieri and G. Moccia, "The grown and characterization of $Hg_{1-x}Cd_xTe$ (CMT) on GaAs for optical fibre communication devices," *Prog. Cryst. Growth and Character.* **19**, 63–81 (1989).

88. L. J. Kozlowski, W. E. Tennant, M. Zandian, J. M. Arias, and J. G. Pasko, "SWIR staring FPA performance at room temperature," *Proc. SPIE* **2746**, 93–100 (1996).

89. L. J. Kozlowski, "HgCdTe focal plane arrays for high performance infrared cameras," *Proc. SPIE* **3179**, 200–211 (1997).

90. R. E. DeWames, D. D. Edwall, M. Zandian, L. O. Bubulac, J. G. Pasko, W. E. Tennant, J. M. Arias, and A. D'Souza, "Dark current generating mechanisms in short wavelength infrared photovolatic detectors," *J. Electron. Mater.* **27**, 722–726 (1998).

91. A. Rogalski and R. Ciupa, "Performance limitation of short wavelength infrared InGaAs and HgCdTe photodiodes," *J. Electron. Mater.* **28**, 630–636 (1999).

92. W. E. Tennant, S. Cabelli, and K. Spariosu, "Prospects of uncooled HgCdTe detector technology," *J. Electron. Mater.* **28**, 582–588 (1999).

93. M. Chu, S. Mesropian, S. Terterian, H. K. Gurgenian, and M. Pauli, "Advanced thermoelectrically cooled midwave HgCdTe focal plane arrays," *J. Electron. Mater.* **33**, 609–614 (2004).

94. B. Orsal, R. Alabedra, M. Valenza, G. Lecoy, J. Meslage, and C. Y. Boisrobert, "$Hg_{0.4}Cd_{0.6}Te$ 1.55-μm avalanche photodiode noise analysis in the vicinity of resonant impact ionization connected with the spin-orbit split-off band," *IEEE Trans. Electron. Devices* **ED-35**, 101–107 (1988).

95. J. D. Beck, C. F. Van, M. A. Kinch, and J. E. Robinson, "MWIR HgCdTe avalanche photodiodes," *Proc. SPIE* **4454**, 188–197 (2001).

96. R. S. Hall, N. T. Gordon, J. Giess, J. E. Hails, A. Graham, D. C. Herbert, D. J. Hall, P. Southern, J. W. Cairns, D. J. Lees, and T. Ashley, "Photomultiplication with low excess noise factor in MWIR to optical fiber compatible wavelengths in cooled HgCdTe mesa diodes," *Proc. SPIE* **5783**, 412–423 (2005).

97. C. Verie, F. Raymond, J. Besson, and T. Nguyen Duy, "Bandgap spin-orbit splitting resonance effects in $Hg_{1-x}Cd_xTe$ alloys," *J. Crystal Growth* **59**, 342–346 (1982).

98. R. Alabedra, B. Orsal, G. Lecoy, G. Pichard, J. Meslage, and P. Fragnon, "An $Hg_{0.3}Cd_{0.7}Te$ avalanche photodiode for optical-fiber transmission systems at $\lambda = 1.3$ μm," *IEEE Trans. Electron. Devices* **ED-32**, 1302–1306 (1985).

99. R. B. Bailey, L. J. Kozlowski, J. Chen, D. Q. Bui, K. Vural, D. D. Edwall, R. V. Gil, A. B. Vanderwyck, E. R. Gertner, and M. B. Gubala, "256×256 hybrid HgCdTe infrared focal plane arrays," *IEEE Trans. Electr. Dev.* **38**, 1104–1109 (1991).

100. L. J. Kozlowski, R. B. Bailey, S. A. Cabelli, D. E. Cooper, I. S. Gergis, A. C. Chen, W. V. McLevige, G. L. Bostrup, K. Vural, W. E. Tennant, and P. H. Howard, "High-performance 5-μm 640×480 HgCdTe-on-sapphire focal plane arrays," *Opt. Eng.* **33**, 54–63 (1994).

101. P. R. Norton, M. Bailey, I. Kasai, and P. King, "Thermoelectrically cooled MWIR HgCdTe image sensors," *Proc. SPIE* **2225**, 289–292 (1994).

102. Vigo System S.A. data sheets (2001).

103. W. V. McLevige, G. M. Williams, R. E. DeWames, J. Bajaj, I. A. Gergis, A. H. Vanderwyck, and E. R. Blazejewski, "Variable-area diode data analysis of surface and bulk effects in MWIR HgCdTe/CdTe/sapphire photodetectors," *Semicond. Sci. Technol.* **8**, 946–952 (1993).

104. J. Antoszewski, unpublished results (2000).

105. N. T. Gordon, R. S. Hall, C. L. Jones, C. D. Maxey, N. E. Metcalfe, R. A. Catchpole, and M. White, "MCT infrared detectors with close to radiatively limited performance at 240 K in the 3–5 μm band," *J. Electron. Mater.* **29**, 818–822 (2000).

106. Z. Djuric, Z. Jaksic, D. Randjelowic, T. Dankocic, W. Ehrfeld, and A. Schmidt, "Enhancement of radiative lifetime in semiconductors using photonic crystals," *Infrared Phys. Technol.* **40**, 25–32 (1999).

107. C. L. Jones, B. E. Metthews, D. R. Purdy, and N. E. Metcalfe, "Fabrication and assessment of optically immersed CdHgTe detector arrays," *Semicond. Sci. Technol.* **6**, 110–113 (1991).

108. J. Piotrowski, "Uncooled infrared photodetectors in Poland," *Proc. SPIE* **5957**, 59570K (2005).

109. T. Ashley, A. B. Dean, C. T. Elliott, M. R. Houlton, C. F. McConville, H. A. Tarry, and C. R. Whitehouse, "Multilayer InSb diodes grown by molecular beam epitaxy for near ambient temperature operation," *Proc. SPIE* **1361**, 238–244 (1990).

110. T. Ashley and T. C. Elliott, "Operation and properties of narrow-gap semiconductor devices near room temperature using non-equilibrium techniques," *Semicond. Sci. Technol.* **8**, C99–C105 (1991).

111. C. T. Elliott, "Advanced heterostructures for $In_{1-x}Al_x$Sb and $Hg_{1-x}Cd_x$Te detectors and emitters," *Proc. SPIE* **2744**, 452–462 (1996).

112. C. T. Elliott, "Non-equilibrium devices in HgCdTe," in *Properties of Narrow Gap Cadmium-based Compounds*, EMIS Data Reviews Series No. 10, P. Capper (Ed.), IEE, London, 339–346 (1994).

113. T. Ashley, C. T. Elliott, N. T. Gordon, R. S. Hall, A. D. Johnson, and G. J. Pryce, "Room temperature narrow gap semiconductor diodes as sources and detectors in the 5–10 μm wavelength region," *J. Crystal Growth* **159**, 1100–1103 (1996).

114. C. T. Elliott, "Photoconductive and non-equilibrium devices in HgCdTe and related alloys," in *Infrared Detectors and Emitters: Materials and Devices*, P. Capper and C. T. Elliott (Eds.), Kluwer Academic Publishers, Boston, 279–312 (2001).

115. M. K. Haigh, G. R. Nash, N. T. Gordon, J. Edwards, A. J. Hydes, D. J. Hall, A. Graham, J. Giess, J. E. Hails, and T. Ashley, "Progress in negative luminescent $Hg_{1-x}Cd_x$Te diode arrays," *Proc. SPIE* **5783**, 376–383 (2005).

116. C. T. Elliott, N. T. Gordon, T. J. Phillips, H. Steen, A. M. White, D. J. Wilson, C. L. Jones, C. D. Maxey, and N. E. Metcalfe, "Minimally cooled heterojunction laser heterodyne detectors in metalorganic vapor phase epitaxially grown $Hg_{1-x}Cd_x$Te," *J. Electron. Mater.* **25**, 1146–1150 (1996).

117. G. J. Bowen, I. D. Blenkinsop, R. Catchpole, N. T. Gordon, M. A. Harper, P. C. Haynes, L. Hipwood, C. J. Hollier, C. Jones, D. J. Lees, C. D. Maxey, D. Milner, M. Ordish, T. S. Philips, R. W. Price, C. Shaw, and P. Southern, "HOTEYE: a novel thermal camera using higher operating temperature infrared detectors," *Proc. SPIE* **5783**, 392–400 (2005).

118. Z. F. Ivasiv, F. F. Sizov, and V. V. Tetyorkin, "Noise spectra and dark current investigations in n^+-p-type $Hg_{1-x}Cd_x$Te ($x \approx 0.22$) photodiodes," *Semicond. Phys. Quantum Electron. Optoelectronics* **2**(3), 21–25 (1999).

119. K. Józwikowski, "Numerical modeling of fluctuation phenomena in semiconductor devices," *J. Appl. Phys.* **90**, 1318–1327 (2001).

120. W. Y. Ho, W. K. Fong, C. Surya, K. Y. Tong, L. W. Lu, and W. K. Ge, "Characterization of hot-electron effects on flicker noise in III-V nitride based heterojunctions," *MRS Internet J. Nitride Semicond. Res.*, 4S1, G6.4 (1999).

121. M. K. Ashby, N. T. Gordon, C. T. Elliott, C. L. Jones, C. D. Maxey, L. Hipwood, and R. Catchpole, "Investigation into the source of $1/f$ noise in $Hg_{1-x}Cd_x$Te diodes," *J. Electron. Mater.* **33**, 757–765 (2004).

122. S. Horn, D. Lohrman, P. Norton, K. McCormack, and A. Hutchinson, "Reaching for the sensitivity limits of uncooled and minimally cooled thermal and photon infrared detectors," *Proc. SPIE* **5783**, 401–411 (2005).

Chapter 7

Photoelectromagnetic, Magnetoconcentration, and Dember IR Detectors

Aside from photoconductive detectors and photodiodes, three other junctionless devices have been used for uncooled IR photodetectors: photoelectromagnetic (or PEM) detectors, magnetoconcentration detectors, and Dember effect detectors. Certainly they are niche devices, but they are still in production and used with important applications, including very fast uncooled detectors with long-wavelength IR radiation. The devices have been reviewed by Piotrowski and Rogalski.[1]

7.1 PEM Detectors

The first experiments on the PEM effect were performed with Cu_2O by Kikoin and Noskov in 1934.[2] Nowak's monograph summarizes the results of his investigations on the PEM effect,[3] which have been replicated worldwide during the last 60 years. For a long time, the PEM effect has been used mostly for InSb room temperature detectors in the middle- and far-IR band. However, the uncooled InSb devices with cutoff wavelength at ≈ 7 μm exhibit no response in the 8–14 μm atmospheric window and relatively modest performance in the 3–5 μm window. $Hg_{1-x}Cd_xTe$ and closely related $Hg_{1-x}Zn_xTe$ and $Hg_{1-x}Mn_xTe$ alloys made it possible to optimize performance of PEM detectors at any specific wavelength.[4]

7.1.1 PEM effect

The PEM effect is caused by diffusion of photogenerated carriers due to the photo-induced carrier in-depths concentration gradient and by deflection of electron and hole trajectories in opposite directions by the magnetic field (Fig. 7.1). If the sample ends are open-circuit in the x-direction, a space charge builds up that gives rise to an electric field along the x axis (open-circuit voltage). If the sample ends are short-circuited in the x-direction, a current flows through the shorting circuit

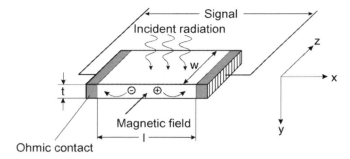

Figure 7.1 Schematic of the PEM effect.

(short-circuit current). In contrast to photoconductors and PV devices, the generation of PEM photovoltage (or photocurrent) requires not simply optical generation, but instead, the formation of an in-depth gradient of photogenerated carriers. Typically, this is accomplished by nonhomogeneous optical generation due to radiation absorption in the near surface region of the device.

7.1.1.1 Transport equations in a magnetic field

Generally, the carrier transport in PEM devices cannot be adequately described by analytical methods and therefore require a numerical solution. Consider the assumptions that make it possible to obtain analytic solutions: homogeneity of the semiconductor, nondegenerate statistics, negligible interface and edge effects, independence of material properties on the magnetic and electric fields, and equal Hall, and drift mobilities. The transport equations for electrons and holes both in x and y take the form of

$$J_{hx} = qp\mu_h E_x + \mu_h B J_{hy}, \tag{7.1}$$

$$J_{hy} = qp\mu_h E_y - \mu_h B J_{hx} - q D_h \frac{dp}{dy}, \tag{7.2}$$

$$J_{ex} = qn\mu_e E_x - \mu_e B J_{ey}, \tag{7.3}$$

and

$$J_{ey} = qn\mu_e E_y + \mu_e B J_{ex} + q D_e \frac{dn}{dy}, \tag{7.4}$$

where B is the magnetic field in the z direction, E_x and E_y are the x and y components of the electric field, J_{ex} and J_{ey} are the x and y components of the electron current density, and J_{hx} and J_{hy} are the analogous components of the hole current density.[5]

E_y can be eliminated in Eqs. (7.1)–(7.4) with the condition

$$J_y = J_{ey} + J_{hy} = 0. \tag{7.5}$$

The other equation to be used is the continuity equation for y-direction currents:

$$\frac{dJ_{hy}}{dy} = -\frac{dJ_{ey}}{dy} = q(G - R),\qquad (7.6)$$

where G and R denote the carrier generation and recombination rates, respectively.

As a result, a nonlinear second-order differential equation for p can be obtained from the set of transport equations for electrons and holes,

$$A_2 \frac{d^2 p}{dy^2} + A_1 \left(\frac{dp}{dy}\right)^2 + A_0 \frac{dp}{dy} - (G - R) = 0,\qquad (7.7)$$

where A_2, A_1, and A_0 are coefficients dependent on the semiconductor parameters and the electric and magnetic fields. Equation (7.7) with the boundary conditions for the front-side and back-side surfaces determines the hole distribution in the y-direction. The electron concentration can be calculated from the electric quasi-neutrality equations. Consequently, the x-direction currents and the electric fields can be calculated.

7.1.1.2 Lile solution

Lile reported the analytical solution for the small-signal, steady-state PEM photovoltage.[5] The voltage responsivity of the PEM detector that can be derived from the Lile solution[4, 6, 7] is

$$R_v = \frac{\lambda}{hc} \frac{B}{wt} \frac{\alpha z(b + 1)}{n_i(b + z^2)} \frac{Z(1 - r_1)}{Y(a^2 - \alpha^2)},\qquad (7.8)$$

where $b = \mu_e/\mu_h$, $z = p/n_i$, w and t are the width and thickness of the detector, r_1 is the front reflectance, and a is the reciprocal diffusion length in the magnetic field equal to

$$a = \left[\frac{(1 - \mu_e^2 B^2)z^2 + b(1 + \mu_h^2 B^2)}{L_e^2(z^2 + 1)}\right]^{1/2}.\qquad (7.9)$$

Z and Y are

$$Z = \left[(\alpha - s_2 a^2 \tau) + (\alpha + s_1 a^2 \tau)\exp(\alpha t)\right]\cosh(-\alpha t) - a\left[(1 - s_2 \alpha t) - (1 + s_1 \alpha t)\right.$$
$$\left. \times \exp(-\alpha t)\right]\sinh(at) - \left[(\alpha + s_1 a^2 \tau) + (\alpha - s_2 a^2 \tau)\exp(-\alpha t)\right],\qquad (7.10)$$

$$Y = a\left[a\tau(s_1 + s_2)\cosh(at) + (1 + s_1 s_2 a^2 \tau^2)\sinh(at)\right]$$
$$- \frac{(b + 1)^2 z^2 \mu_e \mu_h B^2}{a^2(a^2 - \alpha^2)L_e^2 t(b + z^2)(1 + z^2)n_i^2}(a^2 - \alpha^2)n_i^2\{2 - [2 - a^2 t\tau(s_1 + s_1)]$$
$$\times \cosh(at) - a\left[\tau(s_1 + s_2) - t(1 + s_1 s_2 a^2 \tau^2)\right]\sinh(at)\},\qquad (7.11)$$

where s_1 and s_2 are the surface recombination velocities at the front and back surfaces. If $r_2 > 0$, multiple reflections occur, and the resulting voltage responsivity can be calculated by summarizing the contributions from consecutive radiation passes.

The sheet resistivity of the PEM detector is

$$R = \frac{zl}{qn_i\mu_h(b+z^2)wt}\left[1 - \frac{b(b+1)^2z^2\mu_h^2B^2}{a^2L_e^2(1-z^2)(b+z^2)}\right]^{-1}. \qquad (7.12)$$

Since the PEM detector is not biased, Johnson-Nyquist noise is the only noise of the device:

$$V_j = (4kTR\Delta f)^{1/2}, \qquad (7.13)$$

so the detectivity can be calculated from

$$D^* = \frac{R_v(A\Delta f)^{1/2}}{V_j}. \qquad (7.14)$$

The PEM photovoltage is generated along the length of the detector so that the signal linearly increases with the length of the detector and is independent on the device widths for the same photon flux density. This results in a good responsivity for large-area devices, which is in contrast to conventional junction PV devices.

An analysis of Eq. (7.8) indicates that the maximum voltage responsivities can be reached for lightly doped p-type material in magnetic fields $B \approx 1/\mu_e$.[4, 6–8] The advantage of p-type materials is reduced Auger generation and recombination, increased ambipolar diffusion length, absorption coefficient, and resistance.

As has already been pointed out, the optical absorption in $Hg_{1-x}Cd_xTe$ used for uncooled long- wavelength PEM is weak while the absorption depth is typically larger than the diffusion length. Therefore the radiation is almost uniformly absorbed within the ambipolar diffusion length. This results in a low concentration gradient of excess carriers, and consequently, poor performance of PEM devices. The gradient may be induced by the difference in recombination velocities at the front-side and back-side surfaces. Figure 7.2 shows how a low recombination velocity at the front surface and a high recombination velocity at the back surface are important for PEM detector responsivity. The surface recombination velocity required for ultimate performance is quite high ($\approx 10^4$ m/s). In devices with a large difference of recombination velocities and a thickness shorter than the absorption length, the polarity of signal reverses with a change of illumination direction from the low to the high recombination velocity surface while the responsivity remains almost unchanged.

Figure 7.3 shows the properties of detectivity-optimized room temperature 10.6-μm $Hg_{1-x}Cd_xTe$ PEM detectors as a function of doping.[6–8] The best performance is achieved with p-type material doped to $\approx 2 \times 10^{17}$ cm^{-3}. Due to the

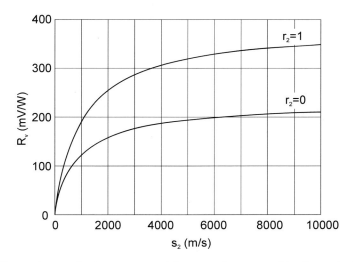

Figure 7.2 Dependence of voltage responsivity on surface recombination velocity for single- and double-pass of radiation The following parameters were used: $\lambda = 10.6$ μm, $T = 300$ K, $N_a = 1.5 \times 10^{17}$ cm^{-3}, $w = 1$ mm, $t = 3$ μm, $r_1 = 0$, and $s_1 = 0$.

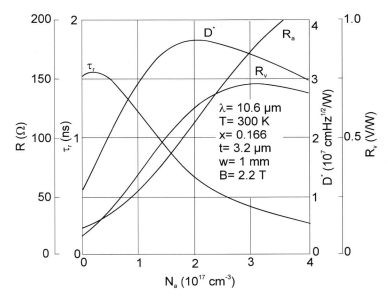

Figure 7.3 Sheet resistivity, voltage responsivity, detectivity, and response time of uncooled 10.6-μm Hg$_{1-x}$Cd$_x$Te PEM detectors as a function of acceptor doping.

high mobility of minority carriers in p-type devices, a magnetic field of ≈2 T is sufficient for good performance.

The voltage responsivity of the detectivity-optimized device is ≈0.6 V/W. The maximum theoretical detectivity of an uncooled 10.6-μm device is ≈3.4 × 10^7 cm Hz$^{1/2}$/W, which is by a factor of ≈3 below the theoretical G-R limit.

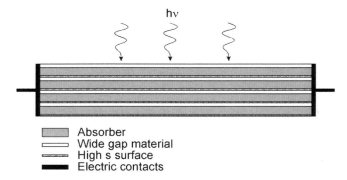

Figure 7.4 Schematic cross-section of the stacked multilayer active element of PEM detectors.

The calculated optimum thickness (≈ 3 µm) is close to the ambipolar diffusion length in a magnetic field. This is less than the optimum thickness for a G-R limited device (see Chapter 2). Therefore, the quantum efficiency and ultimate performance of uncooled LWIR PEM devices is below the G-R limit.

In principle, the performance of PEM detectors could be improved by using stacked multilayer devices composed of thin layers ($t < 1/a$) with high and low surface recombination velocities on opposite surfaces (Fig. 7.4) and the total thickness of absorbing layers close to absorption depths. In such a device, the photogenerated carriers in each absorbing layer can reach the high recombination velocity surface before recombining in volume, and the IR radiation that is not absorbed in top layers can be used for generation in the bottom layers. The individual layers are electrically connected in series or parallel, increasing the resulting photovoltage or photocurrent, respectively.

Cooling to 200 K can increase the detectivity of a PEM detector by 1 order of magnitude. The devices, being relatively bulky, are more difficult to cool than other detectors.

The analysis of PEM detectors prepared from graded gap structures can be made numerically.[9, 10] Although the grading gap structures offer no inherent advantages in terms of ultimate detectivity, the optimized graded gap structures make it possible to obtain good performance without high surface recombination velocity at the backside surface.

Significant improvements in performance can be achieved using optical interference within a device.[1, 3, 4, 8–11] In such a device, a large gradient of optical generation can be achieved even in a weakly absorbing material; as a result, a high recombination velocity is no longer a necessary condition for high response. The spectral response of a resonant cavity PEM detector is highly selective, which may be a disadvantage in some applications.

The response time of a PEM detector may be determined either by the RC time constant or by the decay time of the gradient of excess charge carrier concentration.[4] Typically, the RC time constant of uncooled long wavelength devices is low (<0.1 ns) due to the small capacitance of high-frequency optimized devices ($\approx$1 pF or less) and low resistance ($\approx$50 Ω).

The gradient's decay of the carrier's concentration may be caused by the volume recombination or by ambipolar diffusion. While the first mechanism reduces excess concentration, the second one tends to make the excess concentration uniform. Therefore, the response time can be significantly shorter than the recombination time if the thickness is shorter than the diffusion length. The resulting response time is

$$\frac{1}{\tau_{ef}} = \frac{1}{\tau} + \frac{2D_a}{t^2}. \tag{7.15}$$

For a p-type HgCdTe device with a thickness t of 2 μm, the response time is $\approx$5 $\times$ 10^{-11} s. Even shorter response times are achieved with thinner layers, but at the cost of radiation absorption, quantum efficiency, and detectivity.

7.1.2 Practical HgCdTe PEM detectors

A bulk HgCdTe-based PEM detector was demonstrated for the first time by Giriat and Grynberg.[12] Later this was significantly improved by the use of epitaxial layers.[4, 6–11] The devices are now commercially available.[13]

7.1.2.1 Fabrication

Only epitaxial devices are manufactured at present.[8, 14, 15] The preparation of PEM detectors is essentially very similar to that of photoconductive ones—with the exception of the back surface preparation. In this case, the PEM device is subjected to a special mechanochemical treatment to achieve a high recombination velocity. This procedure is not mandatory when using graded gap structures. Electrical contacts are usually made by Au/Cr deposition, and gold wires are then attached. Figure 7.5 shows the cross-section of a sensitive element of a PEM detector.

Figure 7.6 shows schematically the housings of PEM detectors, which are based on standard TO-5, or for larger elements, on TO-8 transistor cans. Frequently, the

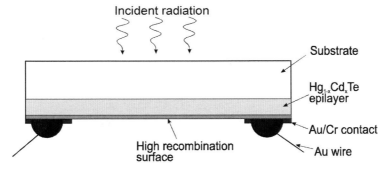

Figure 7.5 Cross-section of a backside-illuminated sensitive element of a PEM detector.

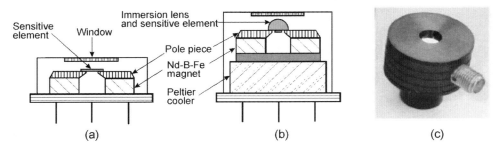

Figure 7.6 Schematic of (a) the housing of an ambient temperature PEM detector; (b) a thermoelectrically cooled, optically immersed PEM detector; and (c) a picture of high-frequency optimized specialized housing. (Reprinted with permission from a Vigo System S.A. data sheet.)

PEM detectors are accommodated in housing dedicated for high-frequency operation. The active elements are mounted in a housing that incorporates a miniature two-element permanent magnet and pole pieces. Magnetic fields approaching 2 T are achievable with the use of modern rare-earth magnetic materials for the permanent magnet and cobalt steel for the pole pieces.

7.1.2.2 Measured performance

Figure 7.7 shows measured spectral detectivity of an uncooled epitaxial $Hg_{1-x}Cd_xTe$ optimized for 10.6-µm operation. The best devices exhibit measured voltage responsivity exceeding 0.15 V/W (width of 1 mm) and detectivities of 1.8×10^7 cm Hz$^{1/2}$/W, a factor of approximately 2 below the predicted ultimate value.[4] The reason for this is a lower (than optimum) magnetic field and faults

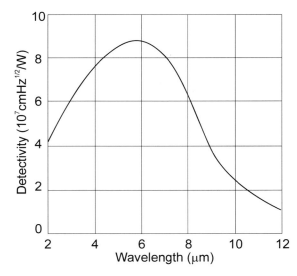

Figure 7.7 Measured spectral detectivity of uncooled 10.6-µm $Hg_{1-x}Cd_xTe$ PEM detectors.

in detector construction, which are probably nonoptimum surface processing and material composition/doping profile.

The fast response of PEM detectors has been confirmed by observations of CO_2 laser self-mode-locking[4] and free-electron laser experiments.[16] When detecting ordinary or low-repetition-rate short pulses, signal voltages up to ≈ 1 V are obtained with 1-mm-long detectors, but they are limited by strong optical excitation effects. The maximum signal voltages for chopped CO_2 radiation are much lower due to radiation heating. In good heat-dissipation design devices, they exceed 30 mV per mm.

Both the theoretical and measured performances of PEM detectors are inferior to those of PC detectors. However, PEM detectors have additional important advantages that make them useful in many applications. In contrast to photoconductors, they do not require electric bias. The frequency characteristics of PEM detector are flat over a wide frequency range, starting from zero frequency. This is due to a lack of low-frequency noise and a very short response time. The resistance of PEM detectors does not decrease with increasing size, which makes it possible to achieve the same performance for small- and large-area devices. With a resistance typically close to 50 Ω, the devices are conveniently coupled directly to wideband amplifiers.

The performance of PEM detectors can be improved by using an optical resonant cavity and optical immersion.[4] With the use of interference, a gain in both voltage responsivity and detectivity can be achieved by a factor >4. Practical optically-immersed PEM devices with lenses formed in transparent substrates of HgCdTe epilayers, in which the expected immersion gain was achieved, have been reported.[4, 17] The size reduction of active elements makes it possible to achieve a strong magnetic field with small magnets.

PEM detectors also have been fabricated with other HgTe-based ternary alloys: $Hg_{1-x}Zn_xTe$[18] and $Hg_{1-x}Mn_xTe$.[19] However, the use of $Hg_{1-x}Zn_xTe$ and $Hg_{1-x}Mn_xTe$ appears to offer no advantages when compared to $Hg_{1-x}Cd_xTe$ in terms of performance and speed of response.

7.2 Magnetoconcentration Detectors

If a semiconductor plate is electrically biased and placed in a crossed magnetic field, then the spatial distribution of electron-hole pairs along the crystal section deviates from the equilibrium value. This is the so-called magnetoconcentration effect. Such redistribution is efficient in a plate whose thickness is comparable to a charge carrier's ambipolar diffusion length. The Lorentz force deflects electrons and holes drifting into the electric field in the same direction, which results in an increase in carrier concentration at one surface and a decrease at the opposite one, in dependence on the direction of the magnetic and electric fields.

The sample shown in Fig. 7.8 has a small recombination velocity at the frontside surface and a high recombination velocity at the back side. This sample is especially interesting for IR detector and source application: When Lorentz force

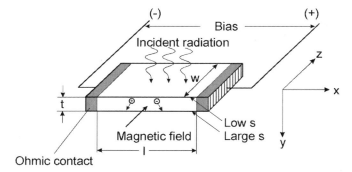

Figure 7.8 Magnetoconcentration effect (depletion mode). Electrons and holes are pushed by action of the Lorentz force toward the back-side surface with a high recombination velocity.

moves carriers toward the high recombination velocity surface (depletion mode), the carriers recombine there, which results in volume depletion with the exception of the region close to the high surface recombination surface. With the opposite direction of the Lorentz force (enrichment mode), the carriers are moved toward the low recombination velocity surface, and are replenished by the generation at the high recombination velocity surface. Changes in the carrier concentration will result in positive or negative luminescence.[20, 21]

Djuric and Piotrowski proposed to use the depletion mode magnetoconcentration effect to suppress the Auger effect.[22–25] They have performed numerical simulation of the magnetoconcetration devices and reported the first practical devices. Theoretical analysis of magnetoconcentration devices can be done only by numerical solution of the set of transport equations for electrons and holes in the presence of a magnetic field [see Eqs. (7.1)–(7.4)]. The steady-state numerical analysis was performed by solving the equations using the fourth-order Runge-Kutta method.

The *I-V* characteristic, differential sheet resistance, and bias power dissipation density as a function of electric field are shown in Fig. 7.9. The *I-V* plot has three characteristic regions. At low electric fields, depletion does not occur, and the plot exhibits ohmic behavior. At higher fields, the plot becomes sub-ohmic, followed by a region of negative resistance as a result of semiconductor depletion. Negative resistance occurs when the carrier concentration decreases faster than drift velocity increases. At a very high field, when the majority carrier concentration saturates at the extrinsic level, the *I-V* plot again becomes linear with a high resistance corresponding to extrinsic concentration.

Figure 7.10 shows calculated parameters of a magnetoconcentration detector as a function of the electric field. The noise current, which at zero electric field is determined by Johnson-Nyquist noise, initially increases with the field, achieving a maximum and then steadily decreasing. The decrease in noise and improvement in detectivity are a result of suppressed Auger-generation noise. Although the depletion and Auger suppression begin at a relatively low electric field ($\approx$40 V/cm), the dependence of the noise current and detectivity on electric field is quite soft; as a result, the saturation at the extrinsic level occurs at much stronger fields

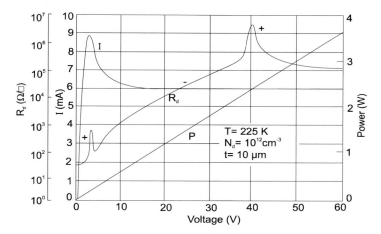

Figure 7.9 The *I-V* characteristics (I), differential sheet resistance (R_d), and bias power dissipation (P) as a function of electric field. The calculations have been performed for a detector size of 1×1 mm^2. The following parameters were used: $x = 0.18$, $N_d = 10^{12}$ cm^{-3}, $T = 225$ K, $t = 15$ μm, $s_1 = 60$ cm/s, and $s_2 = 10^6$ cm/s.

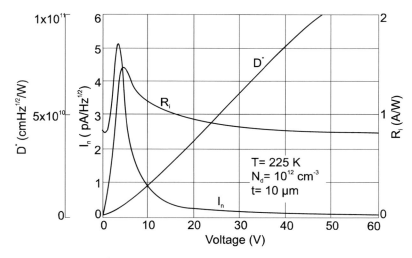

Figure 7.10 The current responsivity (R_i), noise current (I_n), and detectivity (D^*) of a mag-netoconcentration 10.6-μm Hg$_{1-x}$Cd$_x$Te photodetector as a function of the electric field, where $l = w = 1$ mm. The remaining parameters are the same as those defined for Fig. 7.9.

($\approx$600 V/cm). Near-BLIP performance seems to be achievable with two-stage ther-moelectric cooling.

Ohmic contacts may introduce significant thermal generation and may prevent depletion in the nearby region. This may be important for small-sized devices hav-ing a length comparable to the diffusion length.

Practical 10.6-μm uncooled and thermoelectrically cooled magnetoconcentra-tion detectors have been reported.[26] The expected shape of the *I-V* curve with current saturation, negative resistance, and oscillation regions was observed (see

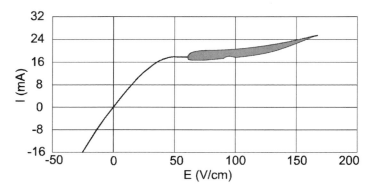

Figure 7.11 Measured current voltage characteristic of an uncooled $Hg_{1-x}Cd_xTe$ 10.6-μm magnetoconcentration detector, where $B = 2$ T and $N_a = 6.2 \times 10^{15}$ cm^{-3}.

Fig. 7.11). The devices exhibited a large low-frequency noise when biased to achieve a sufficient semiconductor depletion. Anomalously high noise has been generated in the region of negative resistance. A significant noise reduction and performance improvement was observed at high frequencies (>100 kHz) with careful selection of a bias current. With such properties, the devices are suitable only for some broadband applications; further efforts are necessary to make them useful in typical applications. Very good heat dissipation is necessary; increasing temperature with bias may result in thermal generation that prevents observing magnetoconcentration depletion.

7.3 Dember Detectors

A Dember detector is a type of PV device that is based on bulk photodiffusion voltage in a simple structure with only one type of semiconductor doping supplied with two contacts.[27] When radiation is incident on the surface of semiconductor-generating electron-hole pairs, a potential difference is usually developed in the direction of the radiation (see Fig. 7.12) as a result of the difference in diffusion of electrons and holes. The Dember effect electrical field restrains the electrons with higher mobility while holes are accelerated, thus making both fluxes equal.

The Dember effect device can be analyzed by solving the transport and continuity equations and by assuming zero total currents in the x and y directions. The steady-state photovoltage under the conditions of weak optical excitation , assuming electroneutrality, can be expressed as

$$V_d = \int_0^t E_z(z)\,dz = \frac{kT}{q}\frac{\mu_e - \mu_h}{n_o\mu_e + p_o\mu_h}[\Delta n(0) - \Delta n(t)], \qquad (7.16)$$

where E_z is the electric field in the z-direction and $\Delta n(z)$ is the excess electron concentration.[28] As this expression shows, two conditions are required for photovoltage generation: the distribution of photogenerated carriers should be nonuni-

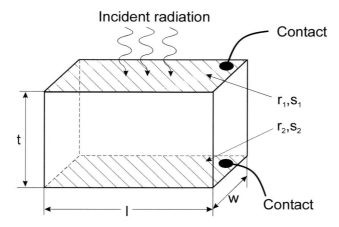

Figure 7.12 Schematic of a Dember detector.

form and the diffusion coefficients of electrons and holes must be different. The gradient may result from nonuniform optical generation and/or from different recombination velocities at the device's front and back surfaces.

For boundary conditions at the top and bottom surface,

$$D_a \frac{\partial \Delta n}{\partial z} = s_1 \Delta n(0) \quad \text{at } z = 0, \tag{7.17}$$

and

$$D_a \frac{\partial \Delta n}{\partial z} = s_2 \Delta n(t) \quad \text{at } z = t. \tag{7.18}$$

The voltage responsivity[29] can be expressed as

$$R_v A = C \left\{ -\left[A(\alpha) + r_2 e^{-2\alpha t} A(-\alpha) \right] \sinh\left(\frac{t}{L_a} \right) + \left[B(\alpha) + r_2 e^{-2\alpha t} B(-\alpha) \right] \right.$$

$$\times \left[1 - \cosh\left(\frac{t}{L_a} \right) \right] + \left(1 - e^{-\alpha t} \right)\left(1 - r_2 e^{-2\alpha t} \right)$$

$$\left. + \left(1 - e^{-\alpha t} \right)\left(1 - r_2 e^{-\alpha t} \right) \right\}, \tag{7.19}$$

where

$$C = \frac{\lambda \alpha \tau}{hc} \frac{\mu_e - \mu_h}{n\mu_e + p\mu_h} \frac{(1 - r_1)kT}{q(1 - \alpha^2 L_a^2)(1 - r_1 r_2 e^{-2\alpha t})}, \tag{7.20}$$

$$A(\alpha) = \frac{(\alpha L_a + \Gamma_1)[\sinh(t/L_a) + \Gamma_2 \cosh(t/L_a)] + \Gamma_1(\alpha L_a - \Gamma_2)e^{-\alpha t}}{(\Gamma_1 + \Gamma_2)\cosh(t/L_a) + (1 + \Gamma_1\Gamma_2)\sinh(t/L_a)}, \tag{7.21}$$

$$B(\alpha) = \frac{(\alpha L_a - \Gamma_2)e^{-\alpha t} - (\Gamma_1 + \alpha L_a)[\cosh(t/L_a) + \Gamma_2 \sinh(t/L_a)]}{(\Gamma_1 + \Gamma_2)\cosh(t/L_a) + (1 + \Gamma_1\Gamma_2)\sinh(t/L_a)}, \tag{7.22}$$

$$\Gamma_1 = \frac{s_1 L_a}{D_a}, \tag{7.23}$$

and

$$\Gamma_1 = \frac{s_2 L_a}{D_a}. \tag{7.24}$$

An analysis of Eqs. (7.19)–(7.24) shows that the maximum response can be achieved for

- Optimized p-type doping,
- A low surface recombination velocity and a low reflection coefficient at the illuminated side contact, and
- A large recombination velocity and a large reflection coefficient at the non-illuminated side contact.

Since the device is not biased and the noise voltage is determined by the Johnson-Nyquist thermal noise, the detectivity [see Eq. (7.14)] can be calculated from Eq. (7.19) and the expression for the Johnson-Nyquist noise [see Eq. (7.13)].

The theoretical design of $Hg_{1-x}Cd_x Te$ and practical Dember effect detectors have been reported.[4, 26, 29] The best performance is achievable for a device thickness slightly larger than the ambipolar diffusion length. More thin devices exhibit a low voltage responsivity while more thick devices have excessive resistance and large related Johnson noise.

Figure 7.13 shows the calculated resistivity, detectivity, and bulk recombination time for the uncooled 10.6-μm Dember detector with detectivity-optimized thickness.[4] As in the case of PEM detectors, the best performance is achieved in

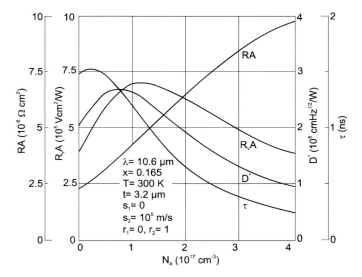

Figure 7.13 The calculated normalized resistivity (RA), normalized responsivity ($R_v A$), detectivity (D^*), and bulk recombination time (τ) of an uncooled 10.6-μm $Hg_{1-x}Cd_x Te$ Dember detector as a function of the acceptor concentration.

p-type material. The calculated detectivity of Dember detectors is comparable to that of photoconductors operated under the same conditions. Detectivities as high as $\approx 2.4 \times 10^8$ cm Hz$^{1/2}$/W and of $\approx 2.2 \times 10^9$ cm Hz$^{1/2}$/W are predicted for optimized 10.6-μm devices at 300 K and 200 K, respectively.

An interesting feature of the Dember device is a significant photoelectric gain that is larger with a zero-bias condition. For optimum doping, the gain is approximately 1.7 and increases with decreasing thickness. The gain is caused by ambipolar effects. At zero bias and with a shorter device, the photogenerated electrons may travel several times between contacts before the holes will recombine or diffuse to the backside contact.

Very low resistances, low voltage responsivities, and noise voltages well below the noise level of the best amplifiers pose serious problems in achieving the potential performance. For example, the optimized uncooled 10.6-μm devices with only 7×7 μm^2 size will have a resistance of ≈ 7 Ω, voltage responsivity of ≈ 82 V/W, and noise voltage of ≈ 0.35 nV/Hz$^{1/2}$. The low resistance and the low voltage responsivity are the reasons why simple Dember detectors cannot compete with other types of photodetectors. But there are some potential ways to overcome these difficulties. One is to connect small-area Dember detectors in a series. For example, to obtain reasonable resistance (≈ 50 Ω) in a 1 mm^2 total area sensitive element of an uncooled 10.6-μm detector, it is necessary to divide its area into 400 50 $\times$ 50 μm^2 elements, connected in series with low resistance interconnections, which is feasible with modern fabrication techniques. The other possibility is to use optical immersion. Detectivities as high as $\approx 1.7 \times 10^9$ cm Hz$^{1/2}$/W and 1.5×10^{10} cm Hz$^{1/2}$/W are possible at 300 K and 200 K in optimized optically immersed 10.6-μm devices.[4]

Due to excess carriers' ambipolar diffusion, the response time of a Dember detector is shorter than the bulk recombination time and can be very short for a device whose thickness is much less than the diffusion length. By shortening the p-type doping with some expense in performance, the response time of detectivity optimized devices is approximately 1 ns. The high-frequency optimized Dember detectors should be prepared from heavily doped material and the thickness should be kept low. For example, for detectors with $N_a = 5 \times 10^{17}$ cm^{-3} and $t = 2$ μm, τ is ≈ 50 ps. Another possible limitation of response time, the RC constant, is not important due to a low resistivity and low capacity of the Dember detector.

Faster and higher-performance Dember detectors can be achieved by using the interference effect. The optimized optical resonant cavity for uncooled 10.6-μm Dember detectors can achieve detectivity by a factor of 5 higher than that of non-interference devices.

The need for a contact with a low surface recombination velocity and a low reflection coefficient at the illuminated side is, perhaps, the main obstacle to implementing practical devices. Remote or heterojunction contacts can be used with the expense of increased series resistance and related Johnson noise.

Figure 7.14 is a schematic of a small-sizes, optically-immersed Dember effect detector manufactured by Vigo System S.A.[4] The sensitive element has been prepared from Hg$_{1-x}$Cd$_x$Te graded gap epilayers grown by isothermal vapor phase

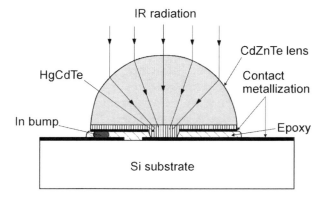

Figure 7.14 Schematic cross-section of an experimental monolithic optically immersed Dember detector.

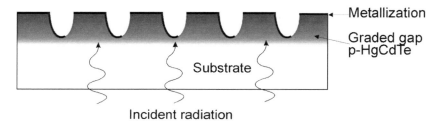

Figure 7.15 Multiple-cell Dember device illuminated from the back side via the substrate.

epitaxy and supplied with a hyperhemispherical lens formed directly in the transparent (Cd,Zn)Te substrate. The surface of the sensitive element was subjected to a special treatment to produce a high recombination velocity and covered with reflecting gold. The "electrical" size of the sensitive element was 7×7 μm^2; due to the hyperhemispherical immersion, the apparent optical size was increased to $\approx 50 \times 50$ μm^2. Multiple cells connected in series Dember detectors (see Fig. 7.15) have been used for large-area devices.

HgCdTe Dember detectors have found application in high-speed laser beam diagnostics of industrial CO_2 lasers and other applications that require fast operational speeds.[30]

References

1. J. Piotrowski and A. Rogalski, "Photoelectromagnetic, magnetoconcentration and Dember infrared detectors," in *Narrow-Gap II-VI Compounds for Optoelectronic and Electromagnetic Applications*, P. Capper (Ed.), Chapman & Hall, London, 507–525 (1997).
2. I. K. Kikoin and M. M. Noskov, "A new photoelectric effect in cooper oxide," *Phys. Z. Sov. Un.* **5**, 586–588 (1934).

3. M. Nowak, "Photoelectromagnetic effect in semiconductors and its applications," *Prog. Quant. Electr.* **11**, 205–346 (1987).

4. J. Piotrowski, W. Galus, and M. Grudzień, "Near room-temperature IR photodetectors," *Infrared Phys.* **31**, 1–48 (1991).

5. D. L. Lile, "Generalized photoelectromagnetic effect in semiconductors," *Phys. Rev. B* **8**, 4708–4722 (1973).

6. D. Genzow, M. Grudzień, and J. Piotrowski, "On the performance of non-cooled CdHgTe photoelectromagnetic detectors for 10.6 μm radiation," *Infrared Phys.* **20**, 133–138 (1980).

7. W. Galus, M. Grudzień, T. Niedziela, T. Persak, and J. Piotrowski, "Photoelectromagnetic CdHgTe detectors for 3–5.5 and 8–14 μm spectral range," *Proc. 9th Int. Symp. of the Technical Committee on Photon Detectors*, IMEKO Secretariat (Ed.), Budapest, 227–233 (1980).

8. J. Piotrowski, "HgCdTe detectors," in *Infrared Photon Detectors*, A. Rogalski (Ed.), SPIE Press, Bellingham, WA, 391–493 (1995).

9. D. Genzow, A. Józwikowska, K. Józwikowski, T. Niedziela, and J. Piotrowski, "Photoelectromagnetic effect (PEM) in graded gap structures (numerical approach)," *Proc. Conf. on Physics of Semiconductor Compounds* **6**, Ossolineum, Wroclaw, 316–320 (1983).

10. D. Genzow, K. Józwikowski, M. Nowak, and J. Piotrowski, "Interference photoelectromagnetic effect in $Hg_{1-x}Cd_xTe$ graded gap structures," *Infrared Phys.* **24**, 21–24 (1984).

11. J. Piotrowski and A. Rogalski, *Semiconductor Infrared Detectors*, WNT, Warsaw (1985) (in Polish).

12. W. Giriat and M. Grynberg, "Photoelectromagnetic IR photodetector," *Przeglad Elektroniki* **5**, 216–221 (1963) (in Polish).

13. Vigo System S.A. data sheets, 2002.

14. J. Piotrowski, "Uncooled operation of IR photodetectors," *Opto-Electr. Rev.* **12**, 11–122 (2004).

15. J. Piotrowski and A. Piotrowski, "Uncooled infrared photodetectors in Poland," *Proc. SPIE* **5957**, 59570K (2005).

16. J. Piotrowski, unpublished results.

17. M. Grudzień and J. Piotrowski, "Monolithic optically immersed HgCdTe IR detectors," *Infrared Phys.* **29**, 251–253 (1989).

18. Z. Nowak, "Technologia, własności elektryczne i fotoelektryczne cienkich warstw $Zn_xHg_{1-x}Te$," doctoral thesis, WAT, Warsaw, 1974 (in Polish).

19. P. Becla, M. Grudzień, and J. Piotrowski, "Uncooled 10.6 μm mercury manganese telluride photoelectromagnetic infrared detectors," *J. Vac. Sci. Technol. B* **9**, 1777–1780 (1991).

20. P. Berdahl, V. Malutenko, and T. Marimoto, "Negative luminescence of semiconductors," *Infrared Phys.* **29**, 667–672 (1989).

21. V. Malyutenko, A. Pigida, and E. Yablonovsky, "Noncooled infrared magnetoinjection emitters based on $Hg_{1-x}Cd_xTe$," *Optoelectronics—Devices and Technologies* **7**, 321–328 (1992).

22. Z. Djuric and J. Piotrowski, "Room temperature IR photodetector with electromagnetic carrier depletion," *Electron. Lett.* **26**, 1689–1691 (1990).

23. Z. Djuric and J. Piotrowski, "Infrared photodetector with electromagnetic carrier depletion," *Opt. Eng.* **31**, 1955–1960 (1992).

24. J. Piotrowski, "Recent advances in IR detector technology," *Microelectr. J.* **23**, 305–313 (1992).

25. Z. Djuric, Z. Jaksic, A. Vujanic, and J. Piotrowski, "Auger generation suppression in narrow-gap semiconductors using the magnetoconcentration effect," *J. Appl. Phys.* **71**, 5706–5708 (1992).

26. J. Piotrowski, W. Gawron, and Z. Djuric, "New generation of near room-temperature photodetectors," *Opt. Eng.* **33**, 1413–1421 (1994).

27. H. Dember, "Uber die Vorwartsbewegung von Elektronen Durch Licht," *Physik Z.* **32**, 554, 856 (1931).

28. J. Auth, D. Genzow, and K. H. Herrmann, *Photoelectrische Erscheinungen*, Akademie Verlag, Berlin (1977).

29. Z. Djuric and J. Piotrowski, "Dember IR photodetectors," *Solid-State Electronics* **34**, 265–269 (1991).

30. H. Heyn, I. Decker, D. Martinen, and H. Wohlfahrt, "Application of room-temperature infrared photo detectors in high-speed laser beam diagnostics of industrial CO_2 lasers," *Proc. SPIE* **2375**, 142–153 (1995).

Chapter 8

Lead Salt Photodetectors

Around 1920, Case investigated the thallium sulfide photoconductor—one of the first photoconductors to give a response in the near IR region to approximately 1.1 μm.[1] The next group of materials to be studied was the lead salts (PbS, PbSe, and PbTe), which extended the wavelength response to 7 μm. PbS photoconductors from natural galena found in Sardinia were originally fabricated by Kutzscher at the University of Berlin in the 1930s.[2] However, for any practical applications it was necessary to develop a technique for producing synthetic crystals. PbS thin-film photoconductors were first produced in Germany, next in the United States at Northwestern University in 1944, and then in England at the Admiralty Research Laboratory in 1945.[3] During World War II, the Germans produced systems that used PbS detectors to detect hot aircraft engines. Immediately after the war, communications, fire control, and search systems began to stimulate a strong development effort that has extended to the present day. After 60 years, low-cost, versatile PbS and PbSe polycrystalline thin films remain the photoconductive detectors of choice for many applications in the 1–3 μm and 3–5 μm spectral range. Current development with lead salts is in the FPAs configuration.

The study of the IV-VI semiconductors received fresh impetus in the mid 1960s with a discovery at Lincoln Laboratories:[4, 5] PbTe, SnTe, PbSe, and SnSe form solid solutions in which the energy gap varies continuously through zero so that it is possible to obtain any required small energy gap by selecting the appropriate composition. For 10 years during the late 1960s to the mid 1970s, HgCdTe alloy detectors were in serious competition with IV-VI alloy devices (mainly PbSnTe) for developing photodiodes because of the latter's production and storage problems.[6, 7] The PbSnTe alloy seemed easier to prepare and appeared more stable. However, development of PbSnTe photodiodes was discontinued because the chalcogenides suffered from two significant drawbacks. The first drawback was a high dielectric constant that resulted in a high diode capacitance, and therefore a limited frequency response. For scanning systems under development at that time, this was a serious limitation. However, for the staring imaging systems that use 2D arrays (which are currently under development), this would not be such a significant issue. The second drawback to IV-VI compounds was their very high thermal coefficient of expansion.[8] This limited their applicability in hybrid configurations with Si multiplexers. Today, with the ability to grow these materials on alternative substrates such as Si, this would not be a fundamental limitation either.

This chapter concentrates on ambient-temperature lead salt photodetectors operated in the SWIR and MWIR spectral ranges. PbS and PbSe IR detectors operate well at comparatively elevated temperatures for their spectral range coverage.

8.1 PbS and PbSe Photoconductors

A number of lead salt photoconductive reviews have been published.[3, 9–22] One of the best reviews of development efforts in lead salt detectors was published by Johnson.[19]

8.1.1 Fabrication

Deposition of polycrystalline PbS and PbSe films is described in Sec. 3.6.2. The films are deposited, either over or under plated gold electrodes, and on fused quartz, crystal quartz, single crystal sapphire, glass, various ceramics, single crystal strontium titanate, Irtran II (ZnS), Si, and Ge. The most commonly used substrate materials are fused quartz for ambient operation and single crystal sapphire for detectors used at temperatures below 230 K. The very low thermal expansion coefficient of fused quartz relative to PbS films results in poorer detector performance at lower operating temperatures. Different shapes of substrates are used: flat, cylindrical, or spherical. To obtain higher collection efficiency, detectors may be deposited directly by immersion onto optical materials with high indices of refraction (e.g., into strontium titanate). Lead salts cannot be immersed directly; special optical cements must be used between the film and the optical element.

As was mentioned in Sec. 3.6.2, in order to obtain high-performance detectors, lead chalcogenide films must be sensitized by oxidation, which may be carried out by using additives in the deposition bath, by post-deposition heat treatment in the presence of oxygen, or by chemical oxidation of the film. Unfortunately, in the older literature the additives are seldom identified and are often referred to only as "an ocidant."[12, 23] A more recent paper deals with the effects of H_2O_2 and $K_2S_2O_8$, both in the deposition bath and in the post-deposition treatment.[24] It was found that both treatments increase the resistivity of PbS films. Although the resistivity usually increases during the oxidation, a different behavior also has been observed.[25] A sensitized PbS film may significantly degrade in air without an overcoating. Possible overcoating materials are As_2S_3, CdTe, ZnSe, Al_2O_3, MgF_2, and SiO_2. Vacuum-deposited As_2S_3 has been found to have the best optical, thermal, and mechanical properties, and it has improved the detector performance. The drawback of the As_2S_3 coating is the toxicity of As and its precursors. Overall, however, the electric properties (resistivity, and in particular, detectivity) of lead salt thin films are rather poorly reported, although there are many papers on PbS and PbSe film growth. The effects of annealing and oxidation treatments on detectivity have not been reported accurately either.

To explain the photoconductivity process in thin film lead salt detectors, three theories have been proposed.[14, 26] The first consists of increases in carrier density during illumination, the oxygen is assumed to introduce a trapping state that inhibits recombination. The second mechanism is based primarily on the increase in the mobility of free carriers in the "barrier model." It is assumed that potential barriers are formed during "sensitization" by heating in oxygen, either between the crystallites of the film or between n- and p-regions in nonuniform films. The third model (commonly referred to as the generalized theory of photoconductivity in semiconducting films in general, and for the lead salts in particular) was proposed by Petritz. In the review paper, Bode concluded that the Petritz theory provided a reasonable framework for general use even though the complex mechanisms in lead salts were still unsolved.[14] More recently, Espevik et al. provided significant additional support for the Petritz theory.[23]

PbS and PbSe materials are peculiar because they have a relatively long response time that affects the significant photoconductive gain. It has been suggested that during the sensitization process the films are oxidized, converting the outer surfaces of exposed PbS and PbSe films to PbO or a mixture of $PbO_x S(Se)_{1-x}$, and forming a heterojunction at the surface (which is shown schematically in Fig. 8.1).[27] Oxide heterointerfaces create conditions for trapping minority carriers or separating majority carriers and thereby extending the lifetime of the material. As was mentioned above, without the sensitization (oxidation) step, lead salt materials have very short lifetimes and a low response.

The basic detector fabrication steps are electrode deposition and delineation, active-layer deposition and delineation, passivation overcoating, mounting (cover/window), and lead wire attachment. Figure 8.2 shows a cross-section of a typical PbS detector structure. The entire device is overcoated for environmental protection and sensitivity improvement. Standard packing consists of electrode deposition and placement in a metal case with a window over the top. Wire leads are placed in a groove in a substrate and gold electrodes are usually vacuum-evaporated onto the film using a mask.

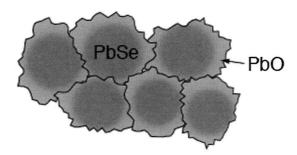

Figure 8.1 PbSe polycrystalline fims after the sensitization process. The films are coated with PbO, forming a heterojunction on the surface. (Reprinted from Ref. 27.)

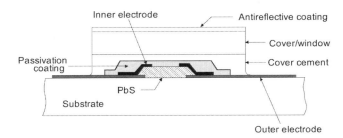

Figure 8.2 Typical structural configuration of PbS detector. (Reprinted from Ref. 19.)

Photolithographic delineation methods are used for complex, high-density patterns of small element sizes. The outer electrodes are produced with vacuum deposition of bimetallic films such as TiAu. To passivate and optimize transmission of radiation into the detector, usually a quarter-wavelength thick overcoating of As_2S_3 is used. Normally, detectors are sealed between a cover plate and the substrate with epoxy cement. The cover-plate material is ordinarily quartz, but other materials such as sapphire may be used to transmit longer wavelengths. This technique seals the detector reasonably well against humid environments. The top surface of the cover is normally made antireflective with a material such as MgF_2.

8.1.2 Performance

Standard active-area sizes are typically 1-, 2-, or 3-mm squares. However, most manufacturers offer sizes ranging from 0.08×0.08 mm^2 to 10×10 mm^2 for PbSe detectors, and 0.025×0.025 mm^2 to 10×10 mm^2 for PbS detectors. Active areas are generally square or rectangular; detectors with more exotic geometries have sometimes not performed up to expectations.

When a lead salt detector is operated below $-20°C$ and exposed to UV radiation, semipermanent changes in responsivity, resistance, and detectivity occur.[21] This is the so-called "flash effect." The amount of change and the degree of permanence depend on the intensity of the UV exposure and the length of exposure time. Lead salt detectors should be protected from fluorescent lighting. They are usually stored in a dark enclosure or overcoated with an appropriate UV-opaque material. They are also hermetically sealed so their long-term stability is not compromised by humidity and corrosion.

The spectral distribution of detectivity of lead salt detectors is presented in Fig. 8.3. Usually the operating temperatures of detectors are between $-196°C$ and $100°C$; it is possible to operate at a temperature higher than recommended, but $150°C$ should never be exceeded. Table 8.1 contains the performance range of detectors fabricated by various manufacturers.

Below 230 K, background radiation begins to limit the detectivity of PbS detectors. This effect becomes more pronounced at 77 K, and peak detectivity is no greater than the value obtained at 193 K. Quantum efficiency of approximately 30% is limited by incomplete absorption of the incident flux in the relatively thin (1- to

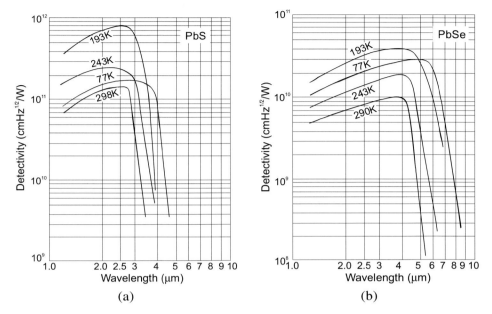

Figure 8.3 Typical spectral detectivity for (a) PbS, and (b) PbSe photoconductors. (Reprinted with permission from a New England Photoconductor (NEP) data sheet.)

Table 8.1 Performance of lead salt detectors (2π FOV, 300 K background). (Reprinted from Ref. 18 with permission from PennWell Corp.)

	T (K)	Spectral response (μm)	λ_p (μm)	D^* (λ_p, 1000 Hz, 1) (cm Hz$^{1/2}$ W^{-1})	$R/\square$ (MΩ)	τ (μs)
PbS	298	1–3	2.5	$(0.1–1.5) \times 10^{11}$	0.1–10	30–1000
	243	1–3.2	2.7	$(0.3–3) \times 10^{11}$	0.2–35	75–3000
	195	1–4	2.9	$(1–3.5) \times 10^{11}$	0.4–100	100–10000
	77	1–4.5	3.4	$(0.5–2.5) \times 10^{11}$	1–1000	500–50000
PbSe	298	1–4.8	4.3	$(0.05–0.8) \times 10^{10}$	0.05–20	0.5–10
	243	1–5	4.5	$(0.15–3) \times 10^{10}$	0.25–120	5–60
	195	1–5.6	4.7	$(0.8–6) \times 10^{10}$	0.4–150	10–100
	77	1–7	5.2	$(0.7–5) \times 10^{10}$	0.5–200	15–150

2-μm) detector material. The responsivity uniformity of PbS and PbSe detectors is generally of the order of 3% to 10%.

The typical frequency response of PbS detectors is shown in Fig. 8.4. An optimum operating frequency arises from the combined effects of the $1/f$ noise at low frequencies and the high-frequency rolloff in responsivity.

Modern lead salt detector arrays contain more than 1000 elements on a single substrate. Operability exceeding 99% is readily achieved for these arrays. Smaller arrays having 100 or fewer elements have been produced with operability of 100%. Arrays as large as several inches on a side use a linear configuration with a single

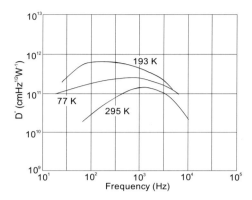

Figure 8.4 Peak spectral detectivity as a function of chopping frequency for a PbS photoconductive detector at 295 K, 193 K, and 77 K operating temperatures. (Reprinted from Ref. 21 with permission from Wiley.)

row and either equal areas equally spaced, or variable sizes. Other configurations include dual rows either in-line or staggered (several staggered rows in staircase fashion, chevron, and double cruciform).

Northrop Grumman EOS coupled 256-pixel PbSe arrays with Si multiplexer readout chips to fabricate assemblies with scanning capabilities.[28] Table 8.2 summarizes the performance of PbS and PbSe arrays in 128- and 256-element configurations. A long-lived thermoelectric element cools the detector/dewar assembly to provide lifetimes greater than 10 years. It should be noted, however, that lead salt photoconductive detectors have a significant $1/f$ noise; e.g., for PbSe, a knee frequency is of the order of 300 Hz at 77 K, 750 Hz at 200 K, and 7 kHz at 300 K.[29] This generally limits the use of these materials to scanning imagers.

Figure 8.5 shows the multimode detector/multiplexer/cooler assembly manufactured by Northrop Grumman. This device consists of a linear or bilinear array of 128 or 256 photoconductive PbSe elements integrated with either a single or dual 128-channel multiplexer chip to give the option of an odd/even or a natural

Table 8.2 Typical performance of PbS and PbSe linear arrays with a CMOS multiplexed readout. Data was taken from Ref. 30.

	PbS		PbSe	
Configuration	128	256	128	256
Element dimensions (μm)	91×102	38×56	91×102	38×56
	(in line)	(staggered)	(in line)	(staggered)
Center spacing (μm)	101.6	50.8	101.6	50.8
D^* (cm Hz$^{1/2}$/W)	3×10^{11}	3×10^{11}	3×10^{10}	3×10^{10}
Responsivity (V/W)	1×10^8	1×10^8	1×10^6	1×10^6
Element time constant (μsec)	≤ 1000	≤ 1000	≤ 20	≤ 20
Nominal element temperature (K)	220	220	220	220
Operability (%)	≥ 98	≥ 98	≥ 98	≥ 98
Dynamic range	$\leq 2000:1$	$\leq 2000:1$	$\leq 2000:1$	$\leq 2000:1$
Channel uniformity (%)	± 10	± 10	± 10	± 10

Figure 8.5 Multimode PbSe FPAs fabricated by Northrop Grumman. (Reprinted from Ref. 31.)

sequence pixel analog output. The multiplexed array is thermoelectrically cooled and closed in a long-life evacuated package with an AR-coated sapphire window and mounted on a circuit board, as shown in the photograph. Similar assemblies containing PbS elements are also fabricated.

In FPA fabrication, lead salt chalcogenides are deposited on Si or SiO from wet chemical baths. Such a monolithic solution avoids the use of a thick slab of these materials mated to Si, as is done with typical hybrids. The detector material is deposited from a wet chemical solution to form polycrystalline photoconductive islands on a CMOS multiplexer. Figure 8.6 shows a few of the 30-μm pixels in this detector array format. Northrop Grumman elaborated on monolithic PbS FPAs in a 320 × 240 format (specified in Table 8.3) with a pixel size of 30 μm.[31] Although PbS photoconductors may be operated satisfactorily at ambient temperature, performance is enhanced by utilizing a self-contained thermoelectric cooler.

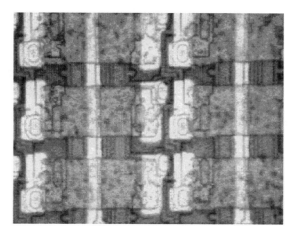

Figure 8.6 Monolithic PbS pixels in a 320 × 240 format. Pixel pitch is 30 μm. (Reprinted from Ref. 31.)

Table 8.3 Specifications for 320 × 240 PbS FPAs.

FPA configuration	Monolithic 320 × 240 PbS
Pixel size (μm)	30 × 30
Detectivity (cm Hz$^{1/2}$/W)	8×10^{10} (ambient); 3×10^{11} (220 K)
Type of signal processor	CMOS
Time constant (ms)	0.2 (ambient); 1 (220 K)
Integration options	Snapshot
Number of output lines	2
Frame rate (Hz)	60
Integration period	Full frame time
Max dynamic range (dB)	69
Active heat dissipation (mW)	200 max
Operability (%)	> 99
Mux transimpedance (MΩ)	100
Detector bias (V)	0–6 (user adjustable)

8.2 Lead Salt Photovoltaic Detectors

In comparison with photoconductive detectors, lead salt photodiodes have been less exploited in the IR region and have not found numerous military and commercial applications. Technological interest in lead salt photodiodes increased in the late 1960s after the discovery of band crossings in the pseudobinary alloys PbSnTe and PbSnSe.[4, 5] However, two drawbacks were identified: (1) the very high thermal coefficients of expansion were not compatible with those of Si by a factor of 7, and (2) a high dielectric constant resulted in limited frequency response, which is a serious limitation for scanning systems. As a result of these drawbacks, the development of IV-VI alloy photodiodes was discontinued in the mid 1970s. But these drawbacks would not present a fundamental limitation today because staring imaging systems currently under development have the ability to grow IV-VI materials on Si through the use of suitable buffer layers. Since work on photovoltaic IV-VI IR detectors is presently restricted to a very small scale, the technical development is far behind that of HgCdTe and related compounds.

Lead salts can exist with very large deviations from steichiometry, and it is difficult to produce doping densities less than 10^{16} cm^{-3} in n-type or 10^{17} cm^{-3} in p-type material. The material becomes n-type under metal-rich conditions and p-type under an excess of chalcogenide. This property is exploited to make photodiodes.

A wide variety of techniques has been used to form p-n junctions in lead salts. They have included interdiffusion, diffusion of donors, ion implantation, proton bombardment, Schottky barrier, and creation of n-type layers on p-type material by VPE or LPE. A summary of works for high-quality photodiode fabrication is given in Table 13.4 of the monograph *Infrared Photon Detectors*.[32]

Historically, interdiffusion was the first technique used to fabricate p-n junctions in lead salts, which produced a change in type due to a change in the stoichimetric defects.[6, 7] Most of the implant work in IV-VI semiconductors has been devoted to the formation of n-on-p junction photodiodes, mainly covering the 3–5 μm spectral region. An alternative technology, adopted especially to prepare long-wavelength photodiodes, was the use of heterojunctions of n-type PbTe (PbSeTe) deposited onto p-type PbSnTe substrates by LPE, VPE, MBE, and hot wall epitaxy (HWE).[32, 33] A considerably simpler technique for photodiode fabrication involves evaporating a thin metal layer onto the semiconductor surface. It is a planar technology with the potential for cheap fabrication of large arrays. An excellent review of Schottky-barrier IV-VI photodiodes with emphasis on thin-film devices has been given by Holloway.[34] Recent progress in the development of lead salt photodiodes has been reviewed by Zogg and Ishida.[22] Table 8.4 summarizes the performance of high-quality lead salt photodiodes operated with near room temperatures.

Detector arrays made from lead salts tend to have very high uniformity in their cutoff wavelengths because of the relatively slow variation in bandgap with composition.

Photodiodes are fabricated in the lead salt layers with metal/semiconductor contacts or p-n junctions. For metal/semiconductor sensors, Pb has proven to be a good blocking material. The p-n junction works well with tellurides. Selenides exhibit a too-high diffusion in order to obtain reliable devices. For both types of junctions, the performance is limited by the dislocation densities in the layers. The low-temperature R_0A products scale linearly with the inverse dislocation density. The optimal carrier concentration in the active region is in the low 10^{17} cm^{-3} range.

A first attempt to realize quasi-monolithic lead salt detector arrays was described by Barrett, Jhabvala, and Maldari, who elaborated on direct integration of PbS photoconductive detectors with MOS transitions.[39, 40] In this process, the PbS films were chemically deposited on the overlaying SiO$_2$ and metallization. Detectivity at 2.0–2.5 μm of 10^{11} cm Hz$^{1/2}$ W^{-1} was measured at 300 K on an integrated photoconductive PbS detector-Si MOSFET preamplifier. Elements of 25×25 μm^2 were easily fabricated.

The research group at the Swiss Federal Institute of Technology demonstrated the first realization of monolithic PbTe FPA (96×128) on a Si substrate containing the active addressing electronics.[41, 42] The monolithic approach used here overcomes the large mismatch in the thermal coefficient of expansion between the group IV-VI materials and Si. Large lattice mismatches between the detector active region and the Si did not impede fabrication of the high-quality layers because the easy plastic deformation of the IV-VIs by dislocation glide on their main glide system without causing structural deterioration.

A schematic cross-section of a PbTe pixel grown epitaxially by MBE on a Si readout structure is shown in Fig. 8.7. A 2–3-nm thick CaF$_2$ buffer layer is employed for compatibility with the Si substrate. Active layers of 2–3-μm thickness only suffice to obtain a near-reflection-loss limited quantum efficiency. The spectral response curves of PbTe photodiodes are shown in Fig. 8.8. Typical quantum efficiencies are around 50% without an AR coating. Metal-semiconductor Pb/PbTe detectors are employed. Each Pb-cathode contact is fed to the drain of the access transistor while the anode (sputtered Pt) is common for all pixels. Figure 8.9 shows a complete array.

Table 8.4 Performance of high-quality lead salt photodiodes operated at ambient temperatures.

Material (concentration)	T (K)	R_oA ($\Omega\,cm^2$)	λ_p (μm)	η_p (%)	D^* ($cm\,Hz^{1/2}\,W^{-1}$)	FOV	Comments	Refs.
$PbSe_{0.8}Te_{0.2}$ ($p \approx 10^{17}\,cm^{-3}$)	170	13–100	3.7–4.1	51–85	$(8-11) \times 10^{11}$	2π	M-S(Pb) values for 3 lateral-collection photodiodes	35
$PbS_{0.63}Se_{0.37}$ ($p \approx 10^{18}\,cm^{-3}$)	195	0.7	3.65	30	9×10^9	90°	Se^+ implant planar	36
	77	5.8×10^3	4.5	36	1.45×10^{11}	90°		
PbS ($p \approx (3 \times 10^{18}\,cm^{-3})$)	300	0.28	2.55	54	4.8×10^9	90°	Se^+ implant planar	37
	195	70	2.95	56	1.1×10^{11}	90°		
	77	7×10^6	3.40	61	6×10^{11}	90°		
PbS ($p \approx 10^{17}\,cm^{-3}$)	300	0.1		50	3×10^9	0	M-S(Pb) Si substrate with CaF_2/BaF_2 buffer, values for a linear 66-element array	38
	200	3		50	2×10^{10}	0		
	84	2×10^5		50	1×10^{13}	0		

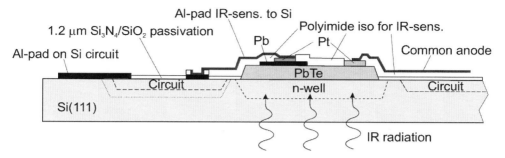

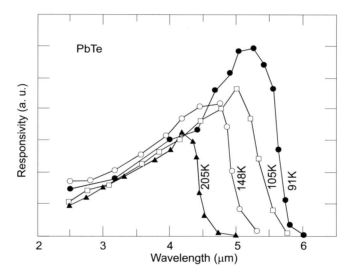

Figure 8.7 Schematic cross-section of one pixel showing the PbTe island as a back-side-illuminated photovoltaic IR detector, the electrical connections to the circuit (access transistor), and a common anode. (Reprinted from Ref. 41 with permission from IEEE.)

Figure 8.8 Spectral responses of a PbTe photodiode. (Reprinted from Ref. 43 with permission from IEEE.)

Despite typical dislocation densities in the 10^7 cm^{-2} range in epitaxial lattice and thermal-expansion-mismatched IV-VI on Si(111) layers, useful photodiodes can be fabricated with $R_o A$ products to 200 Ω cm^2 at 95 K (with a 5.5-μm cutoff wavelength). This is due to the high permittivities of the IV-VI materials, which shield the electric field from charged defects over short distances.

Lead salt heterostructure PV detectors also have been fabricated. The heterojunctions were formed directly between the substrate and the IV-VI films. This solution has been employed to fabricate high-density monolithic PbS-Si heterojunction arrays.[44] However, the performance of heterostructure arrays is inferior to the Schottky barrier and p-n junction arrays.

Figure 8.9 Part of the completely processed monolithic 96 × 128 PbTe-on-Si infrared FPA for the MWIR with the readout electronics in the Si substrate. Pixel pitch is 75 μm. (Reprinted from Ref. 42 with permission from IEEE.)

References

1. T. W. Case, "Notes on the change of resistance of certain substrates in light," *Phys. Rev.* **9**, 305–310 (1917); "The thalofide cell—A new photoelectric substance," *Phys. Rev.* **15**, 289 (1920).

2. E. W. Kutzscher, "Letter to the editor," *Electro-Opt. Syst. Design* **5** (June), 62 (1973).

3. R. J. Cashman, "Film-type infrared photoconductors," *Proc. IRE* **47**, 1471–1475 (1959).

4. J. O. Dimmock, I. Melngailis, and A. J. Strauss, "Band structure and laser action in $Pb_{1-x}Sn_xTe$," *Phys. Rev. Lett.* **16**, 1193–1196 (1966).

5. A. J. Strauss, "Inversion of conduction and valence bands in $Pb_{1-x}Sn_xSe$ alloys," *Phys. Rev.* **157**, 608–611 (1967).

6. J. Melngailis and T. C. Harman, "Single-crystal lead-tin chalcogenides," in *Semiconductors and Semimetals*, Vol. 5, R. K. Willardson and A. C. Beer (Eds.), Academic Press, New York, 111–174 (1970).

7. T. C. Harman and J. Melngailis, "Narrow gap semiconductors," in *Applied Solid State Science*, Vol. 4, R. Wolfe (Ed.), Academic Press, New York, 1–94 (1974).

8. J. T. Longo, D. T. Cheung, A. M. Andrews, C. C. Wang, and J. M. Tracy, "Infrared focal planes in intrinsic semiconductors," *IEEE Trans. Electr. Dev.* **ED-25**, 213–232 (1978).

9. T. S. Moss, "Lead salt photoconductors," *Proc. IRE* **43**, 1869–1881 (1955).

10. P. W. Kruse, L. D. McGlauchlin, and R. B. McQuistan, *Elements of Infrared Technology*, Wiley, New York (1962).

11. D. E. Bode, T. H. Johnson, and B. N. McLean, "Lead selenide detectors for intermediate temperature operation," *Appl. Opt.* **4**, 327–331 (1965).

12. J. N. Humphrey, "Optimization of lead sulfide infrared detectors under diverse operating conditions," *Appl. Opt.* **4**, 665–675 (1965).

13. T. H. Johnson, H. T. Cozine, and B. N. McLean, "Lead selenide detectors for ambient temperature operation," *Appl. Opt.* **4**, 693–696 (1965).

14. D. E. Bode, "Lead salt detectors," in *Physics of Thin Films*, Vol. 3, G. Hass and R. E. Thun (Eds.), Academic Press, New York, 275–301 (1966).

15. T. S. Moss, G. J. Burrel, and B. Ellis, *Semiconductor Optoelectronics*, Butterworths, London (1973).

16. R. E. Harris, "PbS...Mr. Versatility of the detector world," *Electro-Opt. Syst. Design* (now *Laser Focus World*), 47–50 (Dec. 1976).

17. R. E. Harris, "PbSe...Mr Super Sleuth of the detector world," *Electro-Opt. Syst. Design* (now *Laser Focus World*), 42–44 (March 1977).

18. R. E. Harris, "Lead-salt detectors," *Laser Focus/Electro-Optics*, 87–96 (Dec. 1983).

19. T. H. Johnson, "Lead salt detectors and arrays: PbS and PbSe," *Proc. SPIE* **443**, 60–94 (1984).

20. A. Rogalski and J. Piotrowski, "Intrinsic infrared detectors," *Prog. Quant. Electr.* **12**, 87–289 (1988).

21. E. L. Dereniak and G. D. Boreman, *Infrared Detectors and Systems*, Wiley, New York (1996).

22. H. Zogg and A. Ishida, "IV-VI (lead chalcogenide) infrared sensors and lasers," in *Infrared Detectors and Emitters: Materials and Devices*, P. Capper and C. T. Elliott (Eds.), Kluwer, Boston, 43–75 (2001).

23. S. Espevik, C. Wu, and R. H. Bube, "Mechanism of photoconductivity in chemically deposited lead sulfide layers," *J. Appl. Phys.* **42**, 3513–3529 (1971).

24. C. Nascu, V. Vomir, I. Pop, V. Ionescu, and R. Grecu, "The study of PbS films: Influence of oxidants on the chemically deposited PbS thin films," *Mater. Sci. Eng. B* **41**, 235–240 (1996).

25. I. Grozdanov, M. Najdoski, and S. K. Dey, "A simple solution growth technique for PbSe thin films," *Mater. Lett.* **38**, 28–32 (1999).

26. R. H. Bube, *Photoconductivity of Solids*, Wiley, New York (1960).

27. S. Horn, D. Lohrmann, P. Norton, K. McCormack, and A. Hutchinson, "Reaching for the sensitivity limits of uncooled and minimalny-cooled thermal and photon infrared detectors," *Proc. SPIE* **5783**, 401–411 (2005).

28. J. F. Kreider, M. K. Preis, P. C. T. Roberts, L. D. Owen, and W. M. Scott, "Multiplexed mid-wavelength IR long, linear photoconductive focal plane arrays," *Proc. SPIE* **1488**, 376–388 (1991).

29. P. R. Norton, "Infrared image sensors," *Opt. Eng.* **30**, 1649–1663 (1991).

30. Northrup/Grumman Electro-Optical Systems data sheet (2002).

31. T. Beystrum, R. Himoto, N. Jacksen, and M. Sutton, "Low cost PbSalt FPAs," *Proc. SPIE* **5406**, 287–294 (2004).

32. A. Rogalski, "IV-VI detectors," in *Infrared Photon Detectors*, E. Rogalski (Ed.), SPIE Press, Bellingham, WA, 513–559 (1995).

33. A. Rogalski, K. Adamiec, and J. Rutkowski, *Narrow-Gap Semiconductor Photodiodes*, SPIE Press, Bellingham, WA (2000).

34. H. Holloway, "Thin-film IV-VI semiconductor photodiodes," in *Physics of Thin Films*, Vol. 11, G. Haas, M. H. Francombe, and P. W. Hoffman (Eds.), Academic Press, New York, 105–203 (1980).

35. H. Holloway and K. F. Yeung, "Low-capacitance PbTe photodiodes," *Appl. Phys. Lett.* **30**, 210–212 (1977).

36. J. P. Donnelly and T. C. Harman, "P-n junction $PbS_{1-x}Se_x$ photodiodes fabricated by Se^+ ion implantation," *Solid-St. Electr.* **18**, 288–290 (1975).

37. J. P. Donnelly, T. C. Harman, A. G. Foyt, and W. T. Lindley, "PbS photodiodes fabricated by Sb^+ ion implantation," *Solid-St. Electr.* **16**, 529–534 (1973).

38. J. Masek, A. Ishida, H. Zogg, C. Maissen, and S. Blunier, "Monolithic photovoltaic PbS-on-Si infrared-sensor array," *IEEE Electr. Dev. Lett.* **11**, 12–14 (1990).

39. M. D. Jhabvala and J. R. Barrett, "A monolithic lead sulfide-silicon MOS integrated-circuit structure," *IEEE Trans. Electr. Dev.* **ED-29**, 1900–1905 (1982).

40. J. R. Barrett, M. D. Jhabvala, and F. S. Maldari, "Monolithic lead salt-silicon focal plane development," *Proc. SPIE* **409**, 76–88 (1988).

41. K. Alchalabi, D. Zimin, H. Zogg, and W. Buttler, "Monolithic heteroepiraxial PbTe-on-Si infrared focal plane array with 96×128 pixels," *IEEE Electron Dev. Lett.* **22**, 110–112 (2001).

42. H. Zogg, K. Alchalabi, D. Zimin, and K. Kellermann, "Two-dimensional monolithic lead chalcogenide infrared sensor arrays on silicon read-out chips and noise mechanisms," *IEEE Trans. Electron Dev.* **50**, 209–214 (2003).

43. H. Zogg, S. Blunier, T. Hoshino, C. Maissen, J. Masek, and A. N. Tiwari, "Infrared sensor arrays with 3–12 μm cutoff wavelengths in heteroepitaxial narrow-gap semiconductors on silicon substrates," *IEEE Trans. Electr. Dev.* **38**, 1110–1117 (1991).

44. A. J. Steckl, H. Elabd, K. Y. Tam, S. P. Sheu, and M. E. Motamedi, "The optical and detector properties of the PbS-Si heterojunction," *IEEE Trans. Electr. Dev.* **ED-27**, 126–133 (1980).

Chapter 9

Alternative Uncooled Long-Wavelength IR Photodetectors

As alternatives to the current market-dominant HgCdTe, a number of II-VI and III-V semiconductor systems have been proposed, including $Hg_{1-x}Zn_xTe$ (HgZnTe),[1–5] $Hg_{1-x}Mn_xTe$ (HgMnTe),[1–5] $InAs_{1-x}Sb_x$ (InAsSb),[3, 6–8] $InSb_{1-x}Bi_x$ (InSbBi),[9, 10] $In_{1-x}Tl_xSb$ (InTlSb),[10, 11] and InAs-GaSb type-II superlattices.[12, 13]

The wide body of information concerning different methods of crystal growth and physical properties is gathered in Chapter 3. This chapter concentrates on the technology and performance of IR detectors fabricated from Hg-based and III-V materials.

9.1 HgZnTe and HgMnTe Detectors

The technology for HgZnTe IR detectors has benefitted greatly from the HgCdTe device technology base.[1, 3, 5] In comparison with HgCdTe, HgZnTe detectors are easier to prepare due to their relatively higher hardness. The development of device technology requires reproducible high-quality, electronically-stable interfaces with a low interface state density. It was found that the tendency to form surface inversion layers on HgZnTe by anodization is considerably lower than that of HgCdTe.[14] Also, fixed charges at the anodic oxide-HgZnTe interface (2×10^{10} cm^{-2} at 90 K) are lower.[15] Additionally, it has been observed that the anodic oxide-HgZnTe interface is more stable under thermal treatment than the anodic-HgCdTe interface.[16]

The first HgZnTe photoconductive detectors were fabricated by Nowak in the early 1970s.[17] Because of their early stage of development, the performance of these devices was inferior to that of HgCdTe. Then Piotrowski et al. demonstrated that p-type $Hg_{0.885}Zn_{0.115}Te$ can be used as a material for high-quality ambient 10.6-μm photoconductors.[18, 19] These photoconductors, working at 300 K, can achieve 10^8 cm Hz$^{1/2}$ W^{-1} detectivity with optimized composition, doping, and geometry. (Aspects of theoretical performance for both photoconductive and photovoltaic detectors are discussed in Refs. 1–3 and 20–23.)

A HgZnTe ternary alloy also has been used to fabricatePEM detectors. The first nonoptimal detectors were also fabricated by Nowak in the early 1970s.[17] The theoretical design of HgZnTe PEM detectors was reported by Polish workers.[24] An uncooled 10.6-μm HgZnTe PEM detector made it possible to achieve a detectivity value equal to 1.2×10^8 cm Hz$^{1/2}$ W^{-1}.

Several different techniques have yielded p-n HgZnTe junctions, including Hg in-diffusion,[1, 5, 25] Au diffusion,[1, 5, 26] ion implantation,[1, 5, 27–29] and ion etching.[1, 5, 21, 30] To date, the ion implantation method gives the best-quality n$^+$-p HgZnTe photodiodes. The type conversion in p-type material grown by the THM method was achieved by Al implantation.[28, 29]

The HgZnTe photodiode characteristics at 77 K are similar to those of HgCdTe photodiodes. Comparable values of R_0A for both types of photodiodes have been obtained.[3, 28]

Encouraging results have been achieved using a HgCdZnTe quaternary alloy system. Kaiser and Becla produced high-quality p-n junctions from quaternary $Hg_{1-x-y}Cd_xZn_yTe$ epilayers prepared by ISOVPE on CdZnTe substrates.[31] Typical spectral detectivities of p-n HgCdZnTe junctions are presented in Fig. 9.1. The FOV of the diodes was 60°C, with $f = 12$ Hz, and $T_B = 300$ K. Under these conditions, the detectivity of these photodiodes is comparable to those of high-quality HgCdTe photodiodes.

Among the different types of HgMnTe IR detectors, primarily p-n junction photodiodes have been developed.[5, 32, 33] Also, the quaternary HgCdMnTe alloy system is an interesting material for IR applications. The presence of Cd, a third cation in this system, makes it possible to use composition to tune not only the bandgap but also other energy levels, in particular the spin-orbit-split band Γ_7.[34, 35] Because

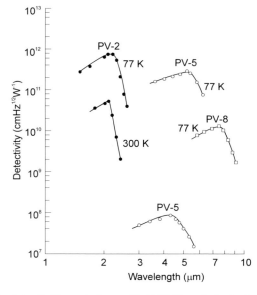

Figure 9.1 Spectral detectivities of three HgCdZnTe photodiodes at 300 and 77 K. (Reprinted from Ref. 31 with permission from Materials Research Society.)

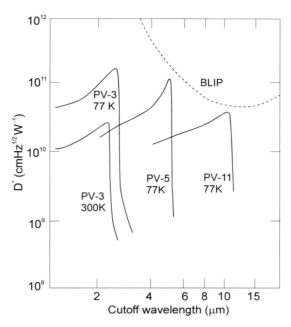

Figure 9.2 Spectral detectivities of HgMnTe and HgCdMnTe photodiodes with 60-deg FOV. (Reprinted from Ref. 32 with permission from AIP.)

of this flexibility, the system appears advantageous, especially for avalanche photodiodes.

Becla produced good-quality p-n HgMnTe and HgCdMnTe junctions by annealing as-grown, p-type samples in Hg-saturated atmospheres.[32] These junctions were made in HgMnTe or HgCdMnTe bulk samples grown by THM, and epitaxial layers grown isothermally on the CdMnTe substrate. Typical spectral detectivities of the HgMnTe and HgCdMnTe photodiodes with 60-deg. FOV are presented in Fig. 9.2, which shows that the detectivities in the 3–5-μm and 8–12-μm spectral ranges are close to the background limit. Typical quantum efficiencies were in the 20–40% range without using an AR coating.

The potential advantage of the HgCdMnTe system is connected with the bandgap spin-orbit-splitting resonance ($E_g = \Delta$) effects in the impact ionization phenomena in short-wavelength avalanche photodiodes (APDs).[36] At room temperature, HgCdTe and HgMnTe systems provide an $E_g = \Delta$ resonance at 1.3 μm and 1.8 μm, respectively. To demonstrate the above possibility of obtaining high-performance HgCdMnTe APDs, Shin et al. used the boron-implantation method to fabricate the mesa-type structures for this quaternary alloy.[37] The R_0A product was 2.62×10^2 Ω cm², which is equivalent to a detectivity value of 1.9×10^{11} cm Hz$^{1/2}$ W^{-1} at 300 K. The breakdown voltage defined at the dark current 10 μA was over 110 V. The leakage current at -10 V was 3×10^{-5} A/cm². This current density is comparable to that reported by Alabedra et al. for the planar bulk HgCdTe APDs.[38] The dark current for the 1.46-μm photodiodes was much lower than the dark currents reported for either bulk or LPE HgCdTe APDs.

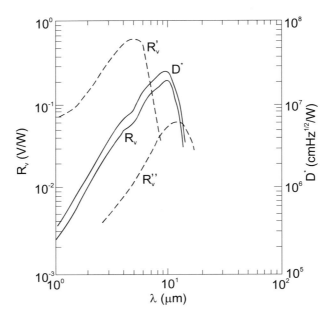

Figure 9.3 Measured spectral responsivity and detectivity of noncooled 10.6-μm $Hg_{1-x}Mn_xTe$ PEM detectors: R_v ($x = 0.08$), R'_v ($x = 0.1$), and R''_v ($x = 0.07$).

Other types of HgMnTe detectors also have been fabricated. Becla et al. described $Hg_{0.92}Mn_{0.08}Te$ PEM detectors with an acceptor concentration of approximately 2×10^{17} cm^{-3}.[39] To obtain such concentrations, the wafers cut from bulk crystals grown by THM were annealed in a Hg-saturated atmosphere. The wafers were then thinned to approximately 5 μm and mounted in the narrow slot of a miniature permanent magnet having a field strength of approximately 1.5 T.

Figure 9.3 shows the experimental spectral voltage responses R_v and detectivity D^* of the HgMnTe PEM detectors.[39] Due to the high level of doping, the spectral characteristics are selective (the spectral half-width is approximately 6 μm). The best performance for detecting IR radiation was achieved using $Hg_{1-x}Mn_xTe$ with composition x of approximately 0.08–0.09. At that composition range, the peak detectivity of the detector was in the region of 7–8 μm. Any reduction or increase in the composition of $Hg_{1-x}Mn_xTe$ leads to a performance reduction at 10.6 μm, as shown in Fig. 9.3.

9.2 III-V Detectors

Earlier data suggest that $InAs_{1-x}Sb_x$ can exhibit a cutoff wavelength up to 12.5 μm at 300 K (the minimum energy gap appears at composition $x = 0.65$).[6] However, some recent experimental results demonstrated that the cutoff wavelength of epitaxial layers can be longer than 12.5 μm; instead, results suggest that it is possible to cover the entire 8–14 μm range at near room temperature. This may be due to structural ordering.[40, 41]

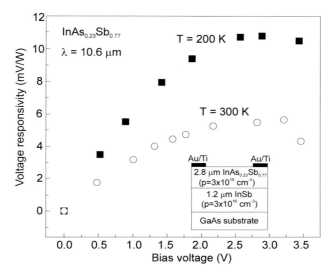

Figure 9.4 Voltage-dependent responsivity of the InAs$_{0.23}$Sb$_{0.77}$ photoconductor at 10.6 μm. (Reprinted from Ref. 7 with permission from Thomson Scientific.)

InAsSb photoconductive detectors are based on p-InAsSb/p-InSb heterostructures. A room-temperature photoresponse up to ≈14 μm has been obtained at 300 K.[7, 41] Figure 9.4 displays the voltage-dependent responsivity at 10.6 μm. It increases with applied voltage and reaches saturation at around 3 V, which corresponds to the values of 5.8 mV/W at 300 K and 10.8 mV/W at 200 K, respectively. From the voltage-dependent responsivity measurement, an effective lifetime of approximately 0.14 ns has been obtained. The estimated detectivity at $\lambda = 10.6$ μm is limited by Johnson noise at the level of approximately 3×10^7 cm Hz$^{1/2}$/W at 300 K. This is below the theoretical limit (≈1.5×10^8 cm Hz$^{1/2}$ W^{-1}) set by the Auger generation-recombination process.

Photovoltaic devices consist of a double heterojunction of p$^+$-InSb/π-InAs$_{1-x}$Sb$_x$/n$^+$-InSb on (001)GaAs [see Fig. 9.5(a)]. In spite of the large lattice mismatch between the InAsSb and GaAs, InAsSb detectors have exhibited good characteristics and showed their feasibility for the near room temperature LWIR photodetectors. The photodiode optimized for $\lambda = 10.6$ μm was characterized by the voltage responsivity-area product of 3×10^{-5} V cm^2/W and detectivity of ≈1.5×10^8 cm Hz$^{1/2}$ W^{-1}. Preliminary experiments on the biased mode operation of InAsSb heterojunction photodiodes demonstrated that the voltage responsivity at near room temperature has been increased by a factor of ≈3 at 0.4 V reverse bias [see Fig. 9.5(b)]. A further increase in responsivity can be achieved through a lower doping level.

Rogalski et al. reported a theoretical analysis of MWIR and LWIR InAs$_{1-x}$Sb$_x$ ($0 \leq x \leq 0.4$) photodiodes with operation extending to the 200–300 K temperature range.[42] It has been shown that the theoretical performance of high-temperature InAsSb photodiodes is comparable to that of HgCdTe photodiodes. Figure 9.6 presents a theoretical limit to the R_0A product for p$^+$-n InAsSb photodiodes op-

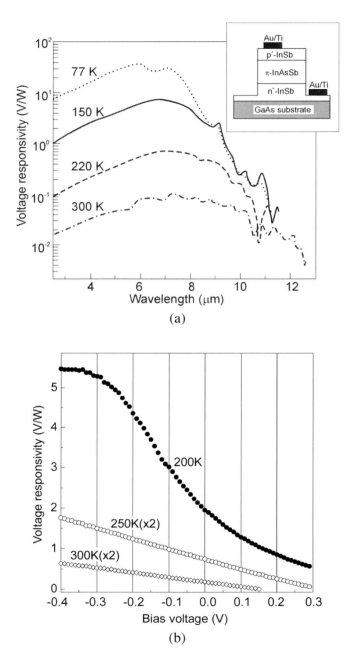

(a)

(b)

Figure 9.5 A p^+-InSb/π-nAs$_{0.15}$Sb$_{0.85}$/n^+-InSb heterojunction photodiode: (a) spectral voltage responsivity (schematic of a photodiode structure in the insert), and (b) bias-dependent voltage responsivity. (Reprinted from Ref. 7 with permission from Thomson Scientific.)

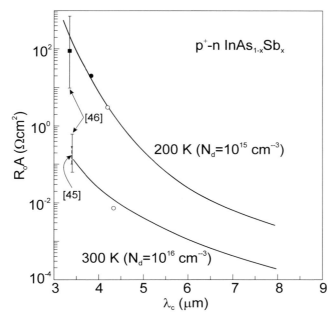

Figure 9.6 The dependence of the R_0A product on the long-wavelength cutoff for p$^+$-n InAs$_{1-x}$Sb$_x$ photodiodes at temperatures 200 and 300 K. The solid curves are calculated for two doping concentrations (10^{15} cm^{-3} and 10^{16} cm^{-3}) in the base region of the photodiode with thickness of 15 µm; a carrier concentration of 10^{18} cm^{-3} in the p$^+$-cap layer with thickness equal to 1 µm is assumed. The experimental data are taken from Refs. 43 ($\bigcirc$), 44 ($\bullet$), 45, and 46.

erated at 200 and 300 K in the 3–8 µm spectral range when doping level in the base n-type layer (10^{15} cm^{-3} and 10^{16} cm^{-3}) is close to the level available in practice.[42] For comparison with theoretical predictions, only limited experimental data is marked. The agreement is satisfactory.

A new material system, InTlSb, was proposed as a potential IR material for the LWIR region.[10, 47] Room-temperature operation of InTlSb photodetectors has been demonstrated with a cutoff wavelength of approximately 11 µm.

As another alternative to the HgCdTe material system, InSb$_{1-x}$Bi$_x$ has been considered.[10, 47] However, the growth of the InSbBi epitaxial layer is difficult due to the large solid-phase miscibility gap between InSb and InBi. Successful growth of the InSbBi epitaxial layer on InSb and GaAs (100) substrates with a substantial amount of Bi ($\sim$ 5%) has been demonstrated using low-pressure MOCVD.[8, 9] The responsivity of the InSb$_{0.95}$Bi$_{0.05}$ photoconductor at 10.6 µm is 1.9×10^{-3} V/W at room temperature, and the corresponding Johnson-noise limited detectivity is 1.2×10^6 cm Hz$^{1/2}$/W.

InSb-based materials also have been used to fabricate nonequilibrium devices (see Sec. 4.3.2). The first nonequilibrium InSb detectors had a p$^+$-π-n$^+$ structure, where π represents low doped p-type material, which is intrinsic at the operating temperature.[46] An accurate analysis of the diode current source at room tempera-

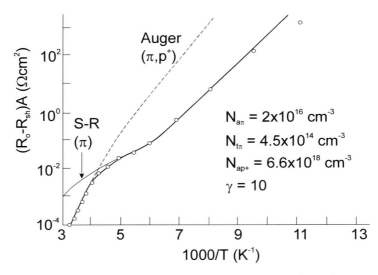

Figure 9.7 Temperature dependence of the R_0A product for a p^+-π-n^+ InSb photodiode. Circles indicate experimental data; the solid lines show calculated values based on both Shockley-Read and Auger-generation mechanisms in the π and p^+ regions. (Reprinted from Ref. 46.)

ture indicates a predominant contribution of A7 generation in the p^+ material. At temperatures below 200 K, diode performance is determined by Shockley-Read generation in the π region (Fig. 9.7). Davies and White's investigatations of the residual currents in Auger-suppressed photodiodes, showed that removing electrons from the active region alters the occupancy of the traps, which leads to an increased generation rate from the Shockley-Read traps in the active region.[48]

It was also shown that a thin strained layer of InAlSb between the p^+ and π regions produces a barrier in the conduction band that substantially reduces the diffusion of electrons from the p^+ layer to the π region, leading to an improvement in room-temperature performance.[49] This type of p^+-P^+-π-n^+ InSb/In$_{1-x}$Sb modified structure is shown schematically in Fig. 9.8. The structure was fabricated by MBE at 420°C using Si-doped (n-type) and Be-doped (p-type) InSb layers. MBE is an attractive growth technology for future devices because it allows accurate control of epilayer purity, thickness, type, and composition, which is coupled with a low-growth temperature that may yield lower Shockley-Read center densities.[50] Typical doping concentrations and thicknesses of individual layers are marked in Fig. 9.8(a). The composition, x, of the In$_{1-x}$Al$_x$Sb barrier is 0.15, which gives a conduction band barrier height estimated to be 0.26 eV. The central region is typically 3 μm thick and not deliberately doped. Circular diodes of 300-μm diameter were fabricated by mesa etching to the p^+ region and were passivated with an anodic oxide. Sputtered Cr/Au contacts were applied to the top of each mesa, with an annual geometry having an internal diameter of 180 μm and an external diameter of 240 μm, and to the p^+ region. A 0.7-μm-thick, thin oxide layer was used as the AR coating.

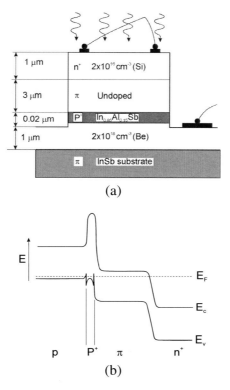

(a)

(b)

Figure 9.8 Schematic cross-section of (a) the p^+-P^+-π-n^+ InSb/In$_{1-x}$Al$_x$Sb heterostructure photodiode, and (b) its energy band diagram. (Reprinted from Ref. 49.)

P^+-P^+-π-n^+ InSb/In$_{1-x}$Al$_x$Sb unbiased heterostructure photodiodes give detectivity above 2×10^9 cm Hz$^{1/2}$ W^{-1} with a peak responsivity at 6 μm. This value is an order of magnitude higher than the value of typical, commercially available, single-element thermal detectors. The structure shown in Fig. 9.8(b) has been tested on 256×256 pixel arrays with a 40-μm pitch. The NEDT in an $f/2.3$ FOV was background-limited up to 110 K.[51]

Figure 9.9 compares theoretical curves for detectivity, calculated from the zero-bias resistance for a conventional p^+-n diode and an epitaxially grown p^+-P^+-ν-n^+ structure with a 3-μm thick active region. There is, for example, an increase in operating temperature of approximately 40 K in the vicinity of 200 K. However, InSb detectors operated at near ambient temperature are not well matched to the 3–5-μm atmospheric transmission window. One solution is to form the active region from In$_{1-x}$Al$_x$Sb with a composition such that the cutoff wavelength is decreased to the optimum value. Obtaining a 5-μm cutoff would require $x \approx 0.023$ at 200 K and $x \approx 0.039$ at room temperature. The predicted detectivity for materials with a constant 5-μm cutoff wavelength is plotted by dotted line in Fig. 9.9. Background-limited detectivity in a 2π FOV can be achieved at 200 K, which is achievable with Peltier coolers, leading to the possibility of high-performance, compact, and comparatively inexpensive imaging systems.

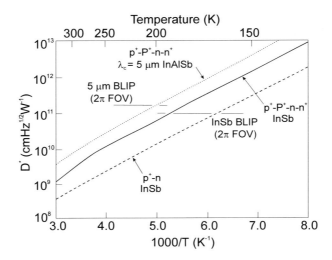

Figure 9.9 Calculated detectivity vs. temperature for InSb photodiodes, comparing bulk InSb (dashed), epitaxial InSb (solid), and epitaxial InAlSb with a 5-μm cutoff wavelength (dotted). (Reprinted from Ref. 49.)

In the first instance, the p^+-n-n^+ 256 × 256 $In_{0.965}Al_{0.035}Sb$ FPAs have been demonstrated. Resonable performance was maintained up to 150 K; at 130 K, the NEDT was below 20 mK.[51]

9.3 InAs/$Ga_{1-x}In_xSb$ Type-II Superlattice Detectors

Type-II superlattices have been proposed as another alternative for IR photodetectors in the LWIR range. In comparison to HgCdTe, the higher effective mass of electrons and holes as well as the slower Auger recombination rate lead to a lower dark current and a higher operating temperature in type-II superlattices. Unlike with type-I superlattices, one can modify the energy of the conduction and valence minibands of a type-II superlattice with a high degree of freedom. The major advantages of the superlattice material are the mechanically robust III-V material and the direct bandgap. InAs-GaInSb type-II superlattices have shown promising results in the LWIR range at low temperatures.[52]

Figure 9.10 compares the R_0A values of an InAs/GaInSb superlattice and HgCdTe photodiodes in the long-wavelength spectral range.[53] The upper line denotes the theoretical diffusion-limited performance corresponding to A7 limitation in p-type HgCdTe material. As can be seen in Fig. 9.10, the most recent photodiode results for superlattice devices rival those of practical HgCdTe devices, thus indicating a substantial improvement in superlattice detector development. Significant improvement in device fabrication has been achieved in the last several years, and the device performance is close to the potential limit.

In the high-temperature range, the performance of LWIR photodiodes is limited by the diffusion process. For example, Fig. 9.11 shows the experimental data and the theoretical prediction of the R_0A product as a function of temperature for

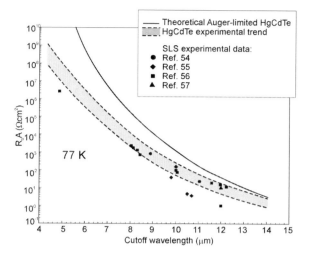

Figure 9.10 Dependence of the R_0A product of InAs/GaInSb SLS photodiodes on cutoff wavelength compared to theoretical and experimental trendlines for comparable HgCdTe photodiodes at 78 K.

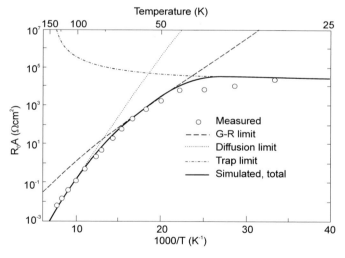

Figure 9.11 The experimental data and the theoretical prediction of the R_0A product as a function of temperature for an InAs/GaInSb photodiode with an 11-μm cutoff wavelength. The activated trap density is taken as a constant (1×10^{12} cm^{-3}) in the simulation over the whole temperature range. (Reprinted from Ref. 54.)

an InAs/GaInSb photodiode with a 11-μm cutoff wavelength. The photodiodes are depletion-region (generation-recombination) limited in the 50–80 K temperature range. The trap-assisted tunneling is dominant only at a low temperature (<50 K) with almost constant activation trap density (1×10^{12} cm^{-3}). Assuming identical material quality with the same activated trap density, the R_0A values exceed 100 Ω cm^2 even though a cutoff wavelength of 14 μm can be achieved.

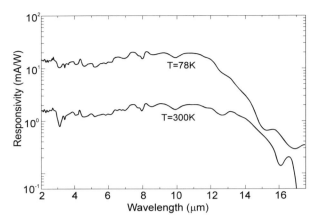

Figure 9.12 The responsivity spectra of the device at 78 and 300 K with an in-plane electrical field of 5 V/cm. (Reprinted from Ref. 12 with permission from IEEE.)

Recently published results also have shown that an $InAs/Ga_{1-x}In_xSb$ type-II material system is very promising for uncooled LWIR detectors.[12, 41, 58] Perhaps the most important problem is still the growth of type-II detectors since the overall performance of the device greatly depends on superlattice quality.

InAs-GaSb superlattices were grown by MBE on semi-insulating (001)GaAs substrates. Photoconductive detectors fabricated from the superlattices showed an 80% cutoff at approximately 12 µm. The responsivity of the device (see Fig. 9.12) is approximately 2 mA/W with a 1 V bias (electrical field 5 V/cm). The maximum measured detectivity of the device is 1.3×10^8 cm Hz$^{1/2}$/W (without any immersion lens or AR coating) at 11 µm at room temperature. The detectivity value is higher than that of the commercially available uncooled HgCdTe detectors at a similar wavelength. The detector shows very weak temperature sensitivity. The carrier lifetime, $\tau = 26$ ns, is an order of magnitude longer than the carrier lifetime in HgCdTe with a similar bandgap and carrier concentration. This provides evidence to the suppression of Auger recombination in this material system.

Figure 9.13 compares the calculated detectivity of Auger generation-recombination limited HgCdTe photodetectors as a function of wavelength and operating temperature with the experimental data of cooled and uncooled type-II detectors at the Center for Quantum Devices, Northwestern University (U.S.). The Auger mechanism is likely to impose fundamental limitations to the LWIR HgCdTe detector performance. The calculations have been performed for doping levels equal to $N_d = 5 \times 10^{15}$ cm^{-3}. These results confirm that the type-II superlattice is a good candidate for IR detectors operating in the spectral range from the mid-wave to the very long-wave IR.

9.4 Thermal Detectors

The use of thermal detectors for IR imaging has been the subject of research and development for many decades. Thermal detectors are not useful for high-speed

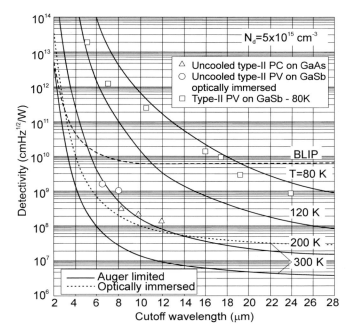

Figure 9.13 Calculated performance of Auger generation-recombination limited HgCdTe photodetectors as a function of wavelength and operating temperature. BLIP detectivity has been calculated for 2π FOV, the background temperature is $T_{BLIP} = 300$ K, and the quantum efficiency $\eta = 1$. The experimental data is taken for cooled and uncooled type-II detectors at the Center for Quantum Devices, Northwestern University (U.S.). The calculations have been performed for a doping level equal to $N_d = 5 \times 10^{15}$ cm^{-3}.

scanning thermal imagers. However, with large arrays of thermal detectors, the best values of NEDT below 0.1 K could be reached because effective noise bandwidths less than 100 Hz can be achieved (see Fig. 9.14).[59–64] Realization of this fact caused a new revolution in thermal imaging, which is underway now. This is due to the development of 2D electronically scanned arrays, in which moderate sensitivity can be compensated with a large number of elements. Large-scale integration combined with micromachining has been used to manufacture large 2D arrays of uncooled IR sensors. This enables fabrication of low-cost and high-quality thermal imagers. Although developed for military applications, low-cost IR imagers are used in nonmilitary applications, such as for a driver's aid, an aircraft aid, industrial process monitoring, community services, firefighting, portable mine detection, night vision, border surveillance, law enforcement, search and rescue, and more.

Figure 9.15 shows a 50-µm pixel in the most common uncooled microbolometer device structure. The microbolometer consists of a 0.5-µm-thick bridge of Si_3N_4 suspended approximately 2 µm above the underlying Si substrate. The use of a vacuum gap of approximately 2.5 µm, together with a quarter-wave resonant cavity between the bolometer and the underlying substrate, can produce a reflector for wavelengths near 10 µm. The bridge is supported by two narrow legs of Si_3N_4,

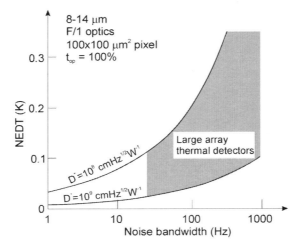

Figure 9.14 The NETD vs. equivalent noise bandwidth for typical thermal detector detectivities. (Reprinted from Ref. 64.)

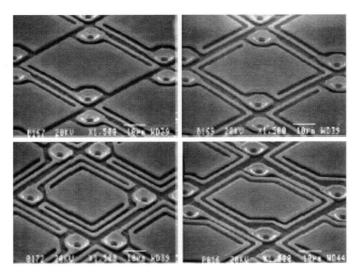

Figure 9.15 Monolithic uncooled microbolometers with 50-μm pixels. Clockwise from upper left: isolation arm on the side, two sides, and twice on two sides. Improving the thermal isolation, which is achieved with long leg lengths, sacrifices the fill factor. (Reprinted from Ref. 65.)

which provide the thermal isolation between the microbolometer and the heat-sink readout substrate. Long, thin legs are anchored to the readout surface. The most popular thermistor materials used to fabricate the monolithic bolometer structures are vanadium dioxide (VO_x) or amorphous Si suspended above the surface in a large resistive pad. Both materials are characterized by high thermal coefficients of resistance, typically about 2%/K.

Figure 9.16 IR image taken by a 480 × 640 pixel microbolometer. These devices can detect temperature variations smaller than 0.1 K. (Reprinted from Ref. 67.)

The detection range of many uncooled IR imaging systems is limited by pixel resolution rather than sensitivity. However, the NEDT is inversely proportional to the pixel area, so the development of highly sensitive 25-μm microbolometer pixels presents significant challenges in both fabrication process improvements and in pixel design. Recent bolometer arrays are only a factor of 2 or 3 away from the temperature fluctuation noise limit for current values of thermal isolation, and optimism remains for further advances.[66] The $f/1$ NEDT performance of 25-μm-pitch microbolometer FPAs is projected to be below 20 mK.[62] Figure 9.16 illustrates an IR image from the large 480 × 640 array format with NEDT below 0.1 K. In general, the NEDT goal is 10 mK for high-performance applications; instead, low-cost performance applications (e.g., security sensors) use 160 × 120 pixels, 50 × 50 μm, and an NEDT of 100 mK.

Pyroelectric detectors are also used for thermal imaging but in a hybrid configuration. The imaging systems that are based on pyroelectric arrays usually must be operated with optical modulators, which chop or defocus the incoming radiation. This may be an important limitation for many applications in which chopperless operation is highly desirable (e.g., guided munitions).

A pyroelectric/ferroelectric bolometer detector [barium strontium titanate (BST) ceramic] was the first hybrid to enter production and is the most widely used type of thermal detector (in the U.S., the Cadillac Division of General Motors has pioneered this application, selling thermal imagers to customers for just under $2000).[61] Although many applications for this hybrid array technology have been identified and imagers employing these arrays are in mass production, no hybrid technology advances are foreseen; the thermal conductance of the bump bonds is so high that the array NETD ($f/1$ optics) is limited to approximately 50 mK. Therefore, pyroelectric array technology is moving toward monolithic Si microstructure technology. The monolithic process should have fewer steps and shorter cycle time. Most ferroelectrics tend to lose their interesting properties as thickness is reduced. However, some ferroelectric materials seem to maintain their properties better than others. This seems particularly true for lead titanate and related materials, whereas with BST, the material does not hold its properties well in the thin-film form.

The typical cost of cryogenically cooled imagers of around \$50,000 (U.S.) restricts their installation to critical military applications involving operations in complete darkness. The commercial systems (microbolometer imagers, radiometers, and ferroelectric imagers) are derived from military systems that are too costly for widespread use. Imaging radiometers employ linear thermoelelctric arrays operating in the snapshot mode; they are less costly than the TV-rate imaging radiometers that employ microbolometer arrays.[61] As the volume of production increases, the cost of commercial systems will inevitably decrease (see Table 9.1).

The performance requirements of near room temperature photodetectors for thermal cameras are considered in Sec. 1.3. Figure 1.7 shows the detectivity dependence on cutoff wavelength for a photon counter detector thermal imager with a resolution of 0.1 K. Detectivities of 1.9×10^8 cm Hz$^{1/2}$/W, 2.3×10^8 cm Hz$^{1/2}$/W, and 2×10^9 cm Hz$^{1/2}$/W are necessary to obtain NEDT $= 0.1$ K for 10-μm, 9-μm, and 5-μm cutoff wavelength photon counter detectors, respectively. These estimations indicate that the ultimate performance of HgCdTe uncooled photodetectors is not sufficient to achieve the required 0.1 K thermal resolution (which is possible with optical immersion or nonequilibrium operation). A thermal resolution below 0.1 K is achieved for staring thermal imagers containing thermal detector FPAs.

As Eq. (1.8) shows, the thermal resolution improves with an increase in detector area, although. increasing the detector area results in reduced spatial resolution. Hence, a reasonable compromise is necessary between the requirements of high thermal resolution and spatial resolution. Improving the thermal resolution without reducing the spatial resolution may be achieved by the following:

Table 9.1 Approximate costs of commercial uncooled arrays for thermal imagers.

Feature	Cost (U.S. \$)
640×480 pixel, 25×25 μm bolometer arrays	15,000
384×288 pixel, 35×35 μm or 25×25 μm bolometer arrays	4000–5500
320×240 pixel, 50×50 μm bolometer arrays	3500–5000
320×240 pixel, 50×50 μm bolometer arrays for imaging radiometers*	15,000–30,000
120×1 pixel, 50×50 μm thermoelectric arrays for imaging radiometers*	<8000
320×240 pixel, 50×50 μm hybrid ferroelectric bolometer array imagers for driver's vision enhancement	1500–3000
160×120 pixel, 50×50 μm bolometer arrays for thermal imagers	<2000
160×120 pixel, 50×50 μm bolometer arrays for driver's vision enhancement systems	<2000
160×120 pixel, 50×50 μm bolometer arrays for imaging radiometers*	<4000

*The cost of arrays for radiometers is considerably higher and depends on specific performance requirements. Estimates given in table should be treated as approximate.

- An increase in the detector area, combined with a corresponding increase in focal length and the objective aperture;
- An improvement in detector performance; and
- An increase in the number of detectors.

An increase in aperture is undesirable because it increases the size, mass, and price of an IR system. It is more desirable to use a detector with higher detectivity. Another possibility is to apply multi-elemental sensors, which reduces each element bandwidth proportionally to the number of elements for the same frame rate and other parameters.

It is of interest to compare the performance of uncooled photon and thermal detectors in the MWIR ($\lambda = 5$ μm) and LWIR ($\lambda = 10$ μm) spectral range. A recently published paper by Kinch includes this comparison.[68] Figure 9.17 compares the theoretical NEDT of detectors operated at 290 K for $f/1$ optics and a 1 mil pixel size. A N^{+}-π-P^{+} HgCdTe photodiode was chosen as a photon detector, as first

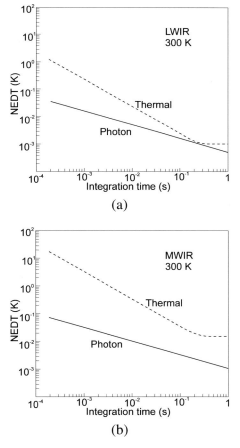

Figure 9.17 Theoretical NEDT comparison of uncooled thermal and HgCdTe uncooled photon (a) LWIR, and (b) MWIR detectors. (Reprinted from Ref. 68 with permission from TMS.)

proposed by Elliott and Ashley,[69] and π designates an intrinsic region containing a p-type background dopant equal to 5×10^{14} cm^{-3} with a carrier lifetime limited by the A7 process. It was also assumed that the detector node capacity could store the integrated charge due to the detector dark current. Figure 9.16 shows that the ultimate performance of uncooled HgCdTe photon detectors is far superior to thermal detectors at wide frame rates and spectral bands.

Comparing both curves for thermal detectors in Fig. 9.16, we see that for long integration times in the LWIR region, excellent performance is achieved when NEDT values are below 10 mK and frame rates are 30 Hz. However, for snapshot systems with integration times below 2 ms, the available NEDT is greater than 100 mK even at the LWIR region. For the MWIR band, the thermal detector has obvious performance limitations at any frame rate.

References

1. A. Rogalski, "Hg$_{1-x}$Zn$_x$Te as a potential infrared detector material," *Prog. Quant. Electr.* **13**, 299–253 (1989).
2. A. Rogalski, "Hg$_{1-x}$Mn$_x$Te as a new infrared detector material," *Infrared Phys.* **31**, 117–166 (1991).
3. A. Rogalski, *New Ternary Alloy Systems for Infrared Detectors*, SPIE Press, Bellingham, WA (1994).
4. J. Piotrowski and A. Rogalski, "Photoelectromagnetic, magnetoconcentration and Dember infrared detectors," in *Narrow-Gap II-VI Compounds for Optoelectronic and Electromagnetic Applications*, P. Capper (Ed.), Chapman & Hall, London, 507–525 (1997).
5. A. Rogalski, "Hg-based alternatives to MCT," in *Infrared Detectors and Emitters: Materials and Devices*, P. Capper and C. T. Elliott (Eds.), Kluwer Academic Publishers, Boston, 377–400 (2001).
6. A. Rogalski, "InAs$_{1-x}$Sb$_x$ infrared detectors," *Prog. Quant. Electron.* **13**, 191–231 (1989).
7. J. D. Kim and M. Razeghi, "Investigation of InAsSb infrared photodetectors for near-room temperature operation," *Opto-Electr. Rev.* **6**, 217–230 (1998).
8. E. Michel and M. Razeghi, "Recent advances in Sb-based materials for uncooled infrared photodetectors," *Opto-Electr. Rev.* **6**, 11–23 (1998).
9. J. L. Lee and M. Razeghi, "Exploration of InSbBi for uncooled long-wavelength infrared photodetectors," *Opto-Electr. Rev.* **6**, 25–36 (1998).
10. J. J. Lee and M. Razeghi, "Novel Sb-based materials for uncooled infrared photodetector applications," *J. Crystal Growth* **221**, 444–449 (2000).
11. H. Asahi, "Tl-based III-V alloy semiconductors," in *Infrared Detectors and Emitters: Materials and Devices*, P. Capper and C. T. Elliott (Eds.), Kluwer Academic Publishers, Boston, 233–349 (2001).
12. H. Mohseni, J. Wojkowski, M. Razeghi, G. Brown, and W. Mitchel, "Uncooled InAs-GaSb type-II infrared detectors grown on GaAs substrates for the 8–12-μm atmospheric window," *IEEE J. Quantum Electron.* **35**, 1041–1044 (1999).

13. M. Razeghi and H. Mohseni, "GaSb/InAs superlattices for infrared FPAs," in *Handbook of Infrared Detection and Technologies*, M. Henini and M. Razeghi (Eds.), Elsevier, Oxford, 191–232 (2002).

14. D. Eger and A. Zigelman, "Anodic oxides on HgZnTe," *Proc. SPIE* **1484**, 48–54 (1991).

15. K. H. Khelland, D. Lemoine, S. Rolland, R. Granger, and R. Triboulet, "Interface properties of passivated HgZnTe," *Semicond. Sci. Technol.* **8**, 56–82 (1993).

16. Yu. V. Medvedev and N. N. Berchenko, "Thermodynamic properties of the native oxide-$Hg_{1-x}Zn_xTe$ interface," *Semicond. Sci. Technol.* **9**, 2253–2257 (1994).

17. Z. Nowak, *Doctoral Thesis*, Military University of Technology, Warsaw (1974) (in Polish).

18. J. Piotrowski, K. Adamiec, A. Maciak, and Z. Nowak, "ZnHgTe as a material for ambient temperature 10.6 μm photodetectors," *Appl. Phys. Lett.* **54**, 143–144 (1989).

19. J. Piotrowski, K. Adamiec, and A. Maciak. "High-temperature 10.6 μm HgZnTe photodetectors," *Infrared Phys.* **29**, 267–270 (1989).

20. J. Piotrowski and T. Niedziela, "Mercury zinc telluride longwavelength high temperature photoconductors," *Infrared Phys.* **30**, 113–119 (1990).

21. A. Rogalski, J. Rutkowski, K. Józwikowski, J. Piotrowski, and Z. Nowak, "The performance of $Hg_{1-x}Zn_xTe$ photodiodes," *Appl. Phys. A* **50**, 379–384 (1990).

22. K. Józwikowski, A. Rogalski, and J. Piotrowski, "On the performance of $Hg_{1-x}Zn_xTe$ photoresistors," *Acta Physica Polonica A* **77**, 359–362 (1990).

23. J. Piotrowski, T. Niedziela, and W. Galus, "High-temperature long-wavelength photoconductors," *Semicond. Sci. Technol.* **5**, S53–S56 (1990).

24. J. Piotrowski, W. Galus, and M. Grudzień, "Near room-temperature IR photodetectors," *Infrared Phys.* **31**, 1–48 (1991).

25. R. Triboulet, T. Le Floch, and J. Saulnier, "First (Hg,Zn)Te infrared detectors," *Proc. SPIE* **659**, 150–152 (1988).

26. Z. Nowak, J. Piotrowski, and J. Rutkowski, "Growth of HgZnTe by cast-recrystallization," *J. Cryst. Growth* **89**, 237–241 (1988).

27. R. Triboulet, "(Hg,Zn)Te: A new material for IR detection," *J. Cryst. Growth* **86**, 79–86 (1988).

28. J. Ameurlaine, A. Rousseau, T. Nguyen-Duy, and R. Triboulet, "(HgZn)Te infrared photovoltaic detectors," *Proc. SPIE* **929**, 14–20 (1988).

29. R. Triboulet, M. Bourdillot, A. Durand, and T. Nguyen Duy, "(Hg,Zn)Te among the other materials for IR detection," *Proc. SPIE* **1106**, 40–47 (1989).

30. P. Brogowski, H. Mucha, and J. Piotrowski, "Modification of mercury cadmium telluride, mercury manganese tellurium, and mercury zinc telluride by ion etching," *Phys. Stat. Sol. (a)* **114**, K37–K40 (1989).

31. D. L. Kaiser and P. Becla, "$Hg_{1-x-y}Cd_xZn_yTe$: Growth, properties and potential for infrared detector applications," *Mat. Res. Soc. Symp. Proc.* **90**, 397–404 (1987).

32. P. Becla, "Infrared photovoltaic detectors utilizing $Hg_{1-x}Mn_xTe$ and $Hg_{1-x-y}Cd_xMn_yTe$ alloys," *J. Vac. Sci. Technol. A* **4**, 2014–2018 (1986).

33. P. Becla, "Advanced infrared photonic devices based on HgMnTe," *Proc. SPIE* **2021**, 22–34 (1993).

34. S. Takeyama and S. Narita, "The band structure parameters determination of the quaternary semimagnetic semiconductor alloy $Hg_{1-x-y}Cd_xMn_yTe$," *J. Phys. Soc. Jap.* **55**, 274–283 (1986).

35. S. Manhas, K. C. Khulbe, D. J. S. Beckett, G. Lamarche, and J. C. Woolley, "Lattice parameters, energy gap, and magnetic properties of the $Cd_xHg_yMn_xTe$ alloy system," *Phys. Stat. Sol.(b)* **143**, 267–274 (1987).

36. R. J. McIntyre, "Multiplication noise in uniform avalanche diodes," *IEEE Trans. Electr. Dev.* **ED-13**, 164–166 (1966).

37. S. H. Shin, J. G. Pasko, D. S. Lo, W. E. Tennant, J. R. Anderson, M. Górska, M. Fotouhi, and C. R. Lu, "$Hg_{1-x-y}Cd_xMn_yTe$ alloys for 1.3–1.8 μm photodiode applications," *Mat. Res. Soc. Symp. Proc.* **89**, 267–274 (1987).

38. R. Alabedra, B. Orsal, G. Lecoy, G. Picard, J. Meslage, and P. Fragnon, "An $Hg_{0.3}Cd_{0.7}Te$ avalanche photodiode for optical-fiber transmission systems at $\lambda = 1.3$ μm," *IEEE Trans. Electr. Dev.* **ED-32**, 1302–1306 (1985).

39. P. Becla, M. Grudzień, and J. Piotrowski, "Uncooled 10.6 μm mercury manganise telluride photoelectromagnetic infrared detectors," *J. Vac. Sci. Technol. B* **9**, 1777–1780 (1991).

40. S. R. Kurtz, L. R. Dawson, R. M. Biefeld, D. M. Follstaedt, and B. L. Doyle, "Ordering-induced band-gap reduction in $InAs_{1-x}Sb$ ($x \approx 0.4$) alloys and superllatices," *Phys. Rev. B* **46**, 1909–1912 (1992).

41. J. D. Kim, D. Wu, J. Wojkowski, J. Piotrowski, J. Xu, and M. Razeghi, "Long-wavelength InAsSb photoconductors operated at near room temperatures (200–300 K)," *Appl. Phys. Lett.* **68**, 99–101 (1996).

42. A. Rogalski, R. Ciupa, and W. Larkowski, "Near room-temperature InAsSb photodiodes: Theoretical predictions and experimental data," *Solid-State Electron.* **39**, 1593–1600 (1996).

43. L. O. Bubulac, E. E. Barrowcliff, W. E. Tennant, J. P. Pasko, G. Williams, A. M. Andrews, D. T. Cheung, and E. R. Gertner, "Be ion implantation in InAsSb and GaInSb," *Inst. Phys. Conf.* Ser. No. 45, 519–529 (1979).

44. M. P. Mikhailova, N. M. Stus, S. V. Slobodchikov, N. V. Zotova, B. A. Matveev, and G. N. Talalakin, "$InAs_{1-x}Sb$ photodiodes for 3–5-μm spectral range," *Fiz. Tekh. Poluprovodn.* **30**, 1613–1619 (1996).

45. EG & G Optoelectronics, Inc., data sheet, 1995.

46. T. Ashley, A. B. Dean, C. T. Elliott, M. R. Houlton, C. F. McConville, H. A. Tarry, and C. R. Whitehouse, "Multilayer InSb diodes grown by molecular beam epitaxy for near ambient temperature operation," *Proc. SPIE* **1361**, 238–244 (1990).

47. M. Razeghi, "Overview of antimonide based III-V semiconductor epitaxial layers and their applications at the Center for Quantum Devices," *Eur. Phys. AP* **23**, 149–205 (2003).

48. A. P. Davis and A. M. White, "Residual noise in Auger suppressed photodiodes," *Infrared Phys.* **31**, 73–79 (1991).

49. C. T. Elliott, "Advanced heterostructures for $In_{1-x}Al_xSb$ and $Hg_{1-x}Cd_xTe$ detectors and emitters," *Proc. SPIE* **2744**, 452–462 (1996).

50. T. Ashley, A. B. Dean, C. T. Elliott, A. D. Johnson, G. J. Pryce, A. M. White, and C. R. Whitehouse, "A heterojunction minority carrier barrier for InSb devices," *Semicond. Sci. Technol.* **8**, S386–S389 (1993).

51. G. R. Nash, N. T. Gordon, D. J. Hall, J. C. Little, G. Masterton, J. E. Hails, J. Giess, L. Haworth, M. T. Emeny, and T. Ashley, "Infrared negative luminescent devices and higher operating temperature detectors," *Proc. SPIE* **5251**, 56–64 (2004).

52. L. Bürkle and F. Fuchs, "InAs/(GaIn)Sb superlattices: a promising material system for infrared detection," in *Handbook of Infrared Detection Technologies*, M. Henini and M. Razeghi (Eds.), Elsevier, Oxford, 159–189 (2002).

53. A. Rogalski and P. Martyniuk, "InAs/GaInSb superlattices as a promising material system for third generation infrared detectors," *Infrared Phys. & Technol.* **48**, 39–52 (2006).

54. O. K. Yang, C. Pfahler, J. Schmitz, W. Pletschen, and F. Fuchs, "Trap centers and minority carrier lifetimes in InAs/GaInSb superlattice long wavelength photodetectors," *Proc. SPIE* **4999**, 448–456 (2003).

55. J. L. Johnson, "The InAs/GaInSb strained layer superlattice as an infrared detector material: An Overview," *Proc. SPIE* **3948**, 118–132 (2000).

56. F. Fuchs, L. Bürkle, R. Hamid, N. Herres, W. Pletschen, R. E. Sah, R. Kiefer, and J. Schmitz, "Optoelectronic properties of photodiodes for the mid- and far-infrared based on the InAs/GaSb/AlSb materials family," *Proc. SPIE* **4288**, 171–182 (2001).

57. M. H. Young, D. H. Chow, A. T. Hunter, and R. H. Miles, "Recent advances in $Ga_{1-x}In_xSb$/InAs superlattice IR detector materials," *Applied Surface Science* **123/124**, 395–399 (1998).

58. J. D. Kim, S. Kim, D. Wu, J. Wojkowski, J. Xu, J. Piotrowski, E. Bigan, and M. Razeghi, "8–13 μm InAsSb heterojunction photodiode operating at near room temperature," *Appl. Phys. Lett.* **67**, 2645–2647 (1995).

59. *Semiconductors and Semimetals*, Vol. 47, P. W. Kruse and D. D. Skatrud (Eds.), Academic Press, San Diego (1997).

60. P. W. Kruse, "Uncooled IR focal plane arrays," *Opto-Electron. Rev.* **7**, 253–258 (1999).

61. P. W. Kruse, *Uncooled Thermal Imaging. Arrays, Systems, and Applications*, SPIE Press, Bellingham, WA (2001).

62. D. Murphy, M. Ray, R. Wyles, J. Asbrock, N. Lum, A. Kennedy, J. Wyles, C. Hewitt, G. Graham, W. Radford, J. Anderson, D. Bradley, R. Chin, and T. Kostrzewa, "High sensitivity (25 μm pitch) microbolometer FPAs and application development," *Proc. SPIE* **4369**, 222–234 (2001).

63. R. Blackwell, S. Geldart, M. Kohin, A. Leary, and R. Murphy, "Recent technology advancements and applications of advanced uncooled imagers," *Proc. SPIE* **5406**, 422–227 (2004).

64. R. Watton and M. V. Mansi, "Performance of a thermal imager employing a hybrid pyroelectric detector array with MOSFET readout," *Proc. SPIE* **865**, 78–85 (1987).

65. P. Norton, J. Campbell, S. Horn, and D. Reago, "Third-generation infrared imagers," *Proc. SPIE* **4130**, 226–236 (2000).

66. C. M. Hanson, "Barriers to background-limited performance for uncooled IR sensors," *Proc. SPIE* **5406**, 454–464 (2004).

67. D. Murphy, M. Ray, R. Wyles, J. Asbrock, N. Lum, J. Wyles, C. Hewitt, A. Kennedy, D. Van Lue, J. Anderson, D. Bradley, R. Chin, and T. Kostrzewa, "High sensitivity 25 μm microbolometer FPAs," *Proc. SPIE* **4820**, 208–219 (2003).

68. M. A. Kinch, "Fundamental physics of infrared detector materials," *J. Electron. Mater.* **29**, 809–817 (2000).

69. T. Ashley and C. T. Elliott, "Non-equilibrium devices for infrared detection," *Electron. Lett.* **21**, 451–452 (1985).

Chapter 10

Final Remarks

10.1 Summary of Progress to Date

During the past several decades, several papers have recognized the fundamental and technological limitations to uncooled photodetector performance.[1-5] Various types of uncooled photodetectors have been developed and their performance has steadily improved. Photoconductors and PEM detectors were the first uncooled devices used for fast detection of long-wavelength radiation without cooling. The present generation of uncooled LWIR devices consists of photovoltaic devices based on $Hg_{1-x}Cd_xTe$ multilayer heterostructures. The problems of poor quantum efficiency and large series resistance have been solved through the adoption of sophisticated heterostructure 3D architectures in combination with reduced physical volume of an active element by using integrated optical concentrators and optical resonance.[6]

Uncooled photodetectors are commercially available and manufactured in significant quantities, mostly as single-element devices.[6, 7] They have found important applications in IR systems that require fast response. Initially, they were mostly CO_2 laser receivers.[1] Examples of present applications are IR spectrophotometers (dispersive, Fourier, and laser), sensitive gas analyzers, nondestructive materials testing, plasma physics, laser metrology and technology, alerters, wide-bandwidth optical communication systems, laser rangefinders, anti-collision systems, smart munition sensors, lidars, and many others.[6] The new applications include various quantum cascade laser-based systems, especially gas analyzers.[8]

Progress in near room temperature FPAs is relatively slow. One possible reason is the more difficult implementation of optical concentrators and low resistance input readout circuits. Recently, Peltier-cooled MWIR focal plane arrays with respectable NETD were demonstrated.[9]

10.2 Directions for Future Development

In general, theoretical analysis has shown that there is no fundamental obstacle to obtaining room temperature operation of photon detectors with background-limited performance and subnanosecond response time.[1–3] The remaining problems are technological rather than fundamental.

Where are we now? The performance of uncooled MWIR photodetectors, and especially thermoelectrically cooled ones, closely approaches the BLIP limit. In contrast, a gap of one order of magnitude still exists between the peak D^* of LWIR detectors (both thermoelectically cooled equilibrium devices and the best uncooled Auger suppressed ones) and the BLIP limit.

Additional gains could be expected with equilibrium devices that are based on known semiconductors since the devices do not reach limits imposed by Auger thermal generation. Significant progress could be expected in uncooled and Peltier-cooled arrays by adopting methods that were previously used for single-element devices and by developing readout circuits optimized for low-resistance detectors. The BLIP performance cannot be achieved in the long-wavelength range with equilibrium-mode $Hg_{1-x}Cd_xTe$ devices operating at temperatures achievable with present Peltier coolers.

The nonequilibrium mode of Auger suppression seems to be a viable approach to drive the operating temperature of the near-BLIP $Hg_{1-x}Cd_xTe$ photodetectors into the uncooled regime.[2,4] Reducing the background acceptor doping below 10^{14} cm^{-3} and trap concentration below $\approx 10^{13}$ cm^{-3} is required for 2π BLIP performance at $\lambda = 10$ μm, as shown in Figs. 10.1 and 10.2. The requirement for doping is softened by almost 1 order of magnitude for 4-μm devices in comparison with 10.6-μm devices. In contrast, trap concentration requirements are more stringent for MWIR devices.

These requirements present a significant material challenge. While donor doping of $\approx 1 \times 10^{14}$ cm^{-3} is possible with background In-doped, Te-rich LPE layers, acceptor doping below 1×10^{15} cm^{-3} is difficult to achieve. The reduction of acceptor concentration is also important because acceptors are expected to act as SR recombination centers.

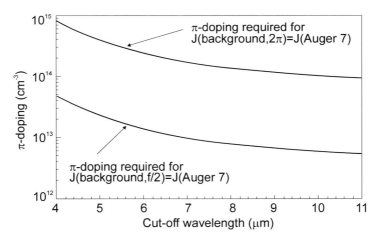

Figure 10.1 Doping at which the A7-generated current is equal to the background-generated levels. (Reprinted from Ref. 2 with permission from AIP.)

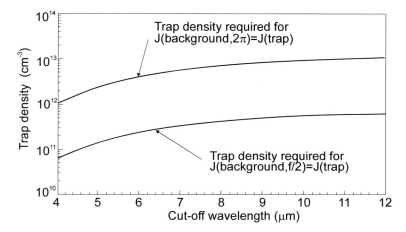

Figure 10.2 Trap concentration at which the trap-generated current is equal to the background-generated levels. (Reprinted from Ref. 2 with permission from AIP.)

In addition to low dopant and trap concentration, a primary unresolved issue is the excess low-frequency noise. Further design refinement of Auger-suppressed devices would be necessary to achieve the potential performance in practice.[10]

It should be noted that the performance of Auger-suppressed photodetectors could be enhanced by the use of optical immersion and a resonant optical cavity. Apart from the usual gain in performance, optical immersion highly reduces the bias power dissipation, which would be important in large-sized elements and in future multielement arrays of photodetectors. Another advantage is the significant reduction of radiative exchange between adjacent elements in an array. These improvements should lead to uncooled near-BLIP photodetectors with present-quality materials.

Another possible solution is to use new natural and artificial semiconductors in which the Auger suppression could be achieved due to the specific electronic structure at the equilibrium mode of operation.[11, 12] They could provide performance at least equivalent to thermal detectors but with cutoff frequencies many orders of magnitude higher.

References

1. J. Piotrowski, W. Galus, and M. Grudzien, "Near room-temperature IR photodetectors," *Infrared Phys.* **31**(1), 1–48 (1991).
2. T. Elliott, N. T. Gordon, and A. M. White, "Towards background-limited, room-temperature, infrared photon detectors in the 3–13 μm wavelength range," *Appl. Phys. Lett.* **74**, 2881–2883 (1999).
3. J. Piotrowski and W. Gawron, "Ultimate performance of infrared photodetectors and figure of merit of detector material," *Infrared Physics and Technology* **38**, 63–68 (1997).

4. M. A. Kinch, "Fundamental physics of infrared detector materials," *J. Electron. Mater.* **29**, 809–817 (2000).

5. M. A. Kinch, "High-operating-temperature (HOT) detector requirements," *Proc. SPIE* **4454**, 168–179 (2001).

6. J. Piotrowski and A. Rogalski, "Uncooled long wavelenth infrared photon detectors," *Infrared Phys. Technol.* **46**, 115–131 (2004).

7. Vigo System S.A., www.vigo.com.pl.

8. D. D. Nelson, B. McManus, S. Urbanski, S. Herndon, and M. S. Zahniser, "High precision measurements of atmospheric nitrous oxide and methane using thermoelectrically cooled mid-infrared quantum cascade lasers and detectors," *Spectrochimica Acta A* **60**, 3325–3335 (2004).

9. G. J. Bowen, I. D. Blenkinsop, R. Catchpole, N. T. Gordon, M. A. Harper, P. C. Haynes, L. Hipwood, C. J. Hollier, C. Jones, D. J. Lees, C. D. Maxey, D. Milner, M. Ordish, T. S. Philips, R. W. Price, C. Shaw, and P. Southern, "HOTEYE: a novel thermal camera using higher operating temperature infrared detectors," *Proc. SPIE* **5783**, 392–400 (2005).

10. S. Horn, D. Lohrman, P. Norton, K. McCormack, and A. Hutchinson, "Reaching for the sensitivity limits of uncooled and minimally cooled thermal and photon infrared detectors," *Proc. SPIE* **5783**, 401–411 (2005).

11. H. Mohseni, J. Wojkowski, M. Razeghi, G. Brown, and W. Mitchel, "Uncooled InAs-GaSb type-II infrared detectors grown on GaAs substrates for the 8–12-μm atmospheric window," *IEEE J. Quantum Electron.* **35**, 1041–1044 (1999).

12. M. Razeghi, "Overview of antimonide based III-V semiconductor epitaxial layers and their applications at the Center for Quantum Devices," *Eur. Phys. AP* **23**, 149–205 (2003).

Index

 **Jozef Piotrowski** is a research and development manager at Vigo System S.A., Warsaw, Poland, and a scientific advisor at the Military Institute of Armament Technology, Zielonka.

He started his career with studies of gigantic photovoltaic effects in semiconductors. By the end of the 1960s, he had recognized the importance of sensitive and fast detection of long-wavelength IR radiation without cooling, and this became his main research subject. He proposed numerous concepts and practical solutions related to uncooled detection. In 1972 he demonstrated uncooled 10-µm photoconductive and photoelectromagnetic detection based on HgCdTe epilayers grown by unique vapor phase deposition on hybrid substrates. During the following decades he developed more advanced IR devices (Dember, magnetoconcentration, and photovoltaic) based on HgCdTe and other material systems (HgZnTe, HgMnTe, InAsSb, InGaAs), various bulk and epitaxial growth techniques (quench/anneal, LPE, MOCVD), and specialized processing techniques (ion milling, ion implantation, photolithography, heterostructure passivation).

In the early 1980s, together with some of his students, Professor Piotrowski founded Vigo System S.A. to commercialize uncooled IR detectors. One of their success stories was the development of low-cost and versatile open-tube gas phase epitaxy, a technique that has been used for commercial detector fabrication for many years. Consequently, in the past decade Professor Piotrowski has introduced into practice various uncooled photodetectors with optical, detection, and electronic functions integrated into a monolithic 3D heterostructure chip. He was also involved in developing advanced detector modules and IR systems based on uncooled photodetectors such as thermal imagers, fast pyrometers, gas analyzers, and IR threat warning, surveillance, and guidance systems. Another field of activity was CdZnTe x-ray and nitride UV detectors. At present, Professor Piotrowski's efforts are concentrated on attaining picosecond response time in the 2–16 µm range with multiple heterojunction IR devices whose performance is close to fundamental limits.

Professor Piotrowski is the author and co-author of about 300 scientific papers, 15 books and monographic papers, and more that 20 patents. Among the honors he has received are the Poland Ministry of Defense Award; the "Photonics Spectra" Circle of Excellence Award (1996) in recognition of excellence, innovation, and achievement in photonics technology; the "Polish Product for Future" Award of the Prime Minister of Poland; foreign membership in the Yugoslav Academy of Engineering; and many others.

Antoni Rogalski, a professor at the Institute of Applied Physics, Military University of Technology in Warsaw, Poland, is one of the world's leading researchers in the field of IR optoelectronics. During the course of his scientific career, he has made pioneering contributions in the theory, design, and technology of different types of IR detectors. In 1997, he received an award from the Foundation for Polish Science, the most prestigious scientific award in Poland, for achievements in the study of ternary alloy systems for IR detectors—mainly alternatives to HgCdTe ternary alloy detectors such as lead salts, InAsSb, HgZnTe, and HgMnTe. In 2004, he was elected as a corresponding member of the Polish Academy of Sciences.

Professor Rogalski's most important scientific achievements include determining the fundamental physical parameters of InAsSb, HgZnTe, HgMnTe, and lead salts; estimating the ultimate performance of ternary alloy detectors; elaborating on studies of high-quality PbSnTe, HgZnTe, and HgCdTe photodiodes operated in the 3–5 and 8–12 µm spectral ranges; and conducting comparative studies of the performance limitations of HgCdTe photodiodes versus other types of photon and thermal detectors (especially QWIR photodetectors).

Professor Rogalski has given more than 35 invited plenary talks at international conferences. He is the author or co-author of approximately 200 scientific papers, 10 books (published by Pergamon Press, SPIE Press, Gordon & Breach, Elsevier, Nauka, and WNT), and 20 book chapters. He is a Fellow of SPIE, the Vice President of the Polish Optoelectronic Committee, a member of the Electronic and Telecommunication Division at the Polish Academy of Sciences, the Editor-in-Chief of the journal *Opto-Electronics Review*, the Deputy Editor-in-Chief of the *Bulletin of the Polish Academy of Sciences: Technical Sciences*, and a member of the editorial boards of the *Journal of Infrared and Millimeter Waves* and the *Journal of Technical Physics*.

Professor Rogalski is also a very active member of the international technical community. He is a co-chair and member of many scientific committees of national and international conferences on optoelectronic devices and crystal growth, the conference chair and organizer of the *International Conference on Solid State Crystals* and the *Material Science and Material Properties for Infrared Optoelectronics* conference, the co-editor of six *SPIE Proceedings* volumes, and a guest editor of *Optical Engineering*.